INSTITUTE OF GEOLOGICAL SCIENCES

Natural Environment Research Council

Mineral Assess

32

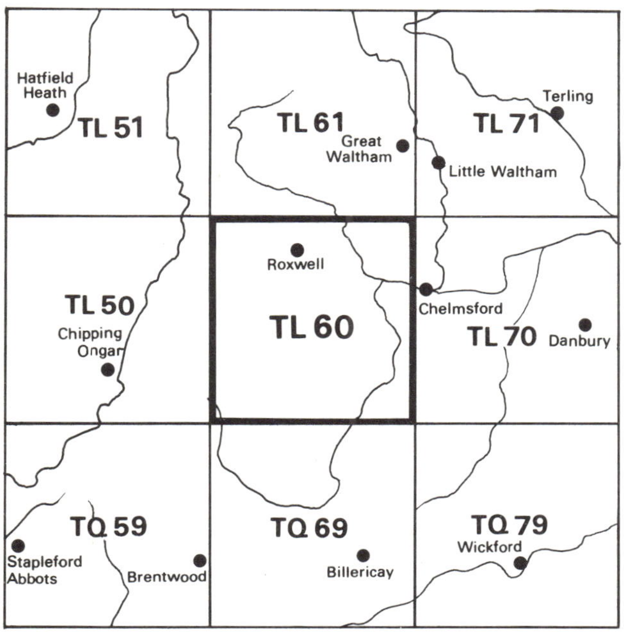

The sand and gravel resources of the country west of Chelmsford, Essex

Description of 1 : 25 000 resource sheet TL 60

P. M. Hopson

ISBN 0 11 884153 X

London Her Majesty's Stationery Office 1981

The first twelve reports on the assessment of British sand and gravel resources appeared in the Report series of the Institute of Geological Sciences as a subseries. Report 13 and subsequent reports appear as Mineral Assessment Reports of the Institute.

Details of published reports appear at the end of this Report.

Any enquiries concerning this report may be addressed to Head, Industrial Minerals Assessment Unit, Institute of Geological Sciences, Keyworth, Nottingham NG12 5GG.

PREFACE

National resources of many industrial minerals may seem so large that stocktaking appears unnecessary, but the demand for minerals and land for all purposes is intensifying and it has become increasingly clear in recent years that regional assessments of the resources of these minerals should be undertaken. The publication of information about the quantity and quality of deposits over large areas is intended to provide a comprehensive factual background against which planning decisions can be made.

Sand and gravel, considered together as naturally occurring aggregate, was selected as the bulk mineral demanding the most urgent attention, initially in the south-east of England, where about half the national output is won and very few sources of alternative aggregates are available. Following a short feasibility project, initiated in 1966 by the Ministry of Land and Natural Resources, the Industrial Minerals Assessment Unit (formerly the Mineral Assessment Unit) began systematic surveys in 1968. The work is now being financed by the Department of the Environment and is being undertaken with the cooperation of the Sand and Gravel Association of Great Britain.

This report describes the resources of sand and gravel of 100 km^2 of the country west of Chelmsford, Essex, shown on the accompanying 1:25 000 resource sheet TL 60. The survey was conducted by Mr P. M. Hopson, assisted by Mr R. J. Marks and Mr D. W. Murray during the drilling and sampling programme. Mr Hopson compiled the report. The work is based on a geological survey at the 1:10 560 scale by Mr R. D. Lake and Mr R. A. Ellison in 1969 and 1977. Mr J. D. Burnell, ISO, (Land Agent) has been responsible for negotiating access to land for drilling. The ready cooperation of land owners and tenants in this work is gratefully acknowledged.

G. M. Brown
Director

Institute of Geological Sciences
Exhibition Road
London SW7 2DE

4 December 1980

CONTENTS

FIGURES

MAP

TABLES

The sand and gravel resources of the country west of Chelmsford, Essex

Description of 1:25 000 resource sheet TL 60

P. M. HOPSON

SUMMARY

The geological maps of the Institute of Geological Sciences, pre-existing borehole information and 42 boreholes drilled for the Industrial Minerals Assessment Unit form the basis of the assessment of the sand and gravel resources of the country west of Chelmsford.

All deposits in the area which might be potentially workable for sand and gravel have been investigated and a simple statistical method has been used to estimate the volume. The reliability of the volume estimates is given at the symmetrical 95 per cent probability level.

The 1:25 000 map is divided into 3 resource blocks containing between 2.1 and 10.5 km^2 of potentially workable sand and gravel. For the blocks assessed statistically the geology of the deposits is described and the mineral-bearing area, the mean thickness of overburden and mineral, and the mean grading of the mineral are stated. Detailed borehole data are given. The geology, the position of the boreholes and the outlines of the resource blocks are shown on the accompanying map.

Bibliographic reference
HOPSON, P. M. 1981. The sand and gravel resources of the country west of Chelmsford, Essex. Description of 1:25 000 resource sheet TL 60. *Miner. Assess. Rep. Inst. Geol. Sci.*, No. 66.

Note
All National Grid references in this report lie in the 1:25 000 sheet TL 60

Author
P. M. Hopson, BSc.
Institute of Geological Sciences
Keyworth, Nottingham NG12 5GG

INTRODUCTION

The survey is concerned with the estimation of resources, which include deposits that are not currently exploitable but have a foreseeable use, rather than reserves, which can only be assessed in the light of current, locally prevailing, economic considerations. Clearly, both the economic and the social factors used to decide whether a deposit may be workable in the future cannot be predicted; they are likely to change with time. Deposits not currently economically workable may be exploited as demand increases, as higher grade or alternative materials become scarce, or as improved processing techniques are applied to them. The improved knowledge of the main physical properties of the resource and their variability which this survey seeks to provide, will add significantly to the factual background against which planning policies can be decided (Archer, 1969; Thurrell, 1971; Harris and others, 1974).

The survey provides information at the 'indicated' level "for which tonnage and grade are computed partly from specific measurements, samples or production data and partly from projection for a reasonable distance on geologic evidence. The sites available for inspection, measurement, and sampling are too widely or otherwise inappropriately spaced to permit the mineral bodies to be outlined completely or the grade established throughout" (Bureau of Mines and Geological Survey, 1948, p. 15).

It follows that the whereabouts of reserves must still be established and their size and quality proved by the customary detailed exploration and evaluation undertaken by the industry. However, the information provided by this survey should assist in the selection of the best targets for such further work. The following arbitrary physical criteria have been adopted:

a The deposit should average at least one metre in thickness.
b The ratio of overburden to sand and gravel should be no more than 3:1.
c The proportion of fines (particles passing the No. 240 mesh BS sieve, about $\frac{1}{16}$ mm) should not exceed 40 per cent.
d The deposit should lie within 25 m of the surface, this being taken as the likely maximum working depth under most circumstances. It follows from the second criterion that boreholes are drilled no deeper than 18 m if no sand and gravel has been proved.

A deposit of sand and gravel which broadly meets these criteria is regarded as 'potentially workable' and is described and assessed as 'mineral' in this report. As the assessment is at the indicated level, parts of such a deposit may not satisfy all the criteria.

For the particular needs of assessing sand and gravel resources, a grain-size classification based on the

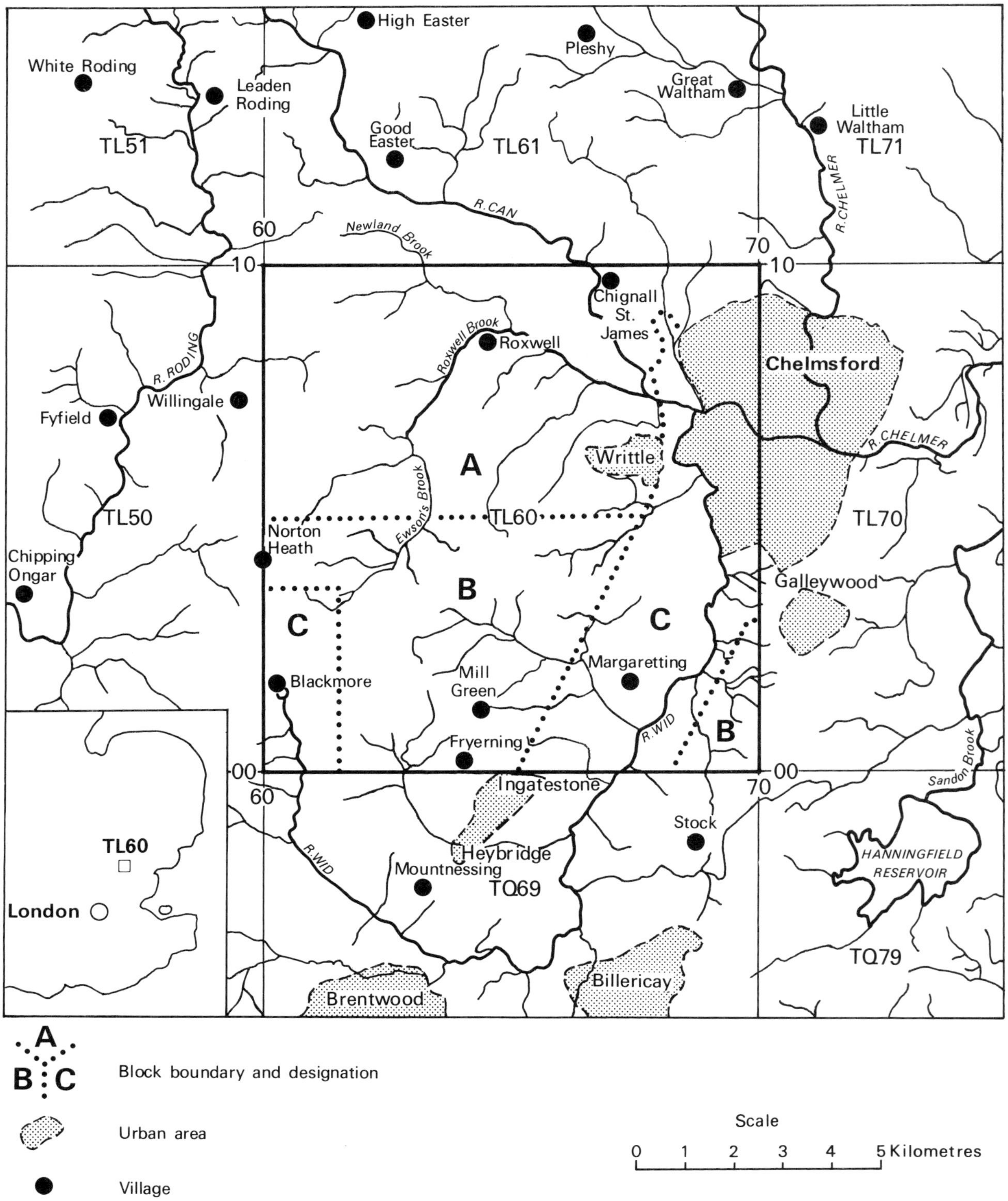

Figure 1 Sketch map of the location of sheet TL 60, showing the resource block boundaries.

geometric scale $\frac{1}{16}$ mm, $\frac{1}{4}$ mm, 1 mm, 4 mm, 16 mm has been adopted. The boundaries between fines (that is, the clay and silt fractions) and sand, and between sand and gravel grade material, are placed at $\frac{1}{16}$ mm and 4 mm respectively (see Appendix C).

The volume and other characteristics are assessed within resource blocks, each of which, ideally, contains approximately 10 km^2 of sand and gravel. No account is taken of any factors, for example, roads, villages and high agricultural or landscape value, which might stand in the way of sand and gravel being exploited, although towns are excluded. The estimated total volume therefore bears no simple relationship to the amount that could be extracted in practice.

It must be emphasised that the assessment applies to the resource block as a whole. Valid conclusions cannot be drawn about the mineral in parts of a block, except in the immediate vicinity of the actual sample points.

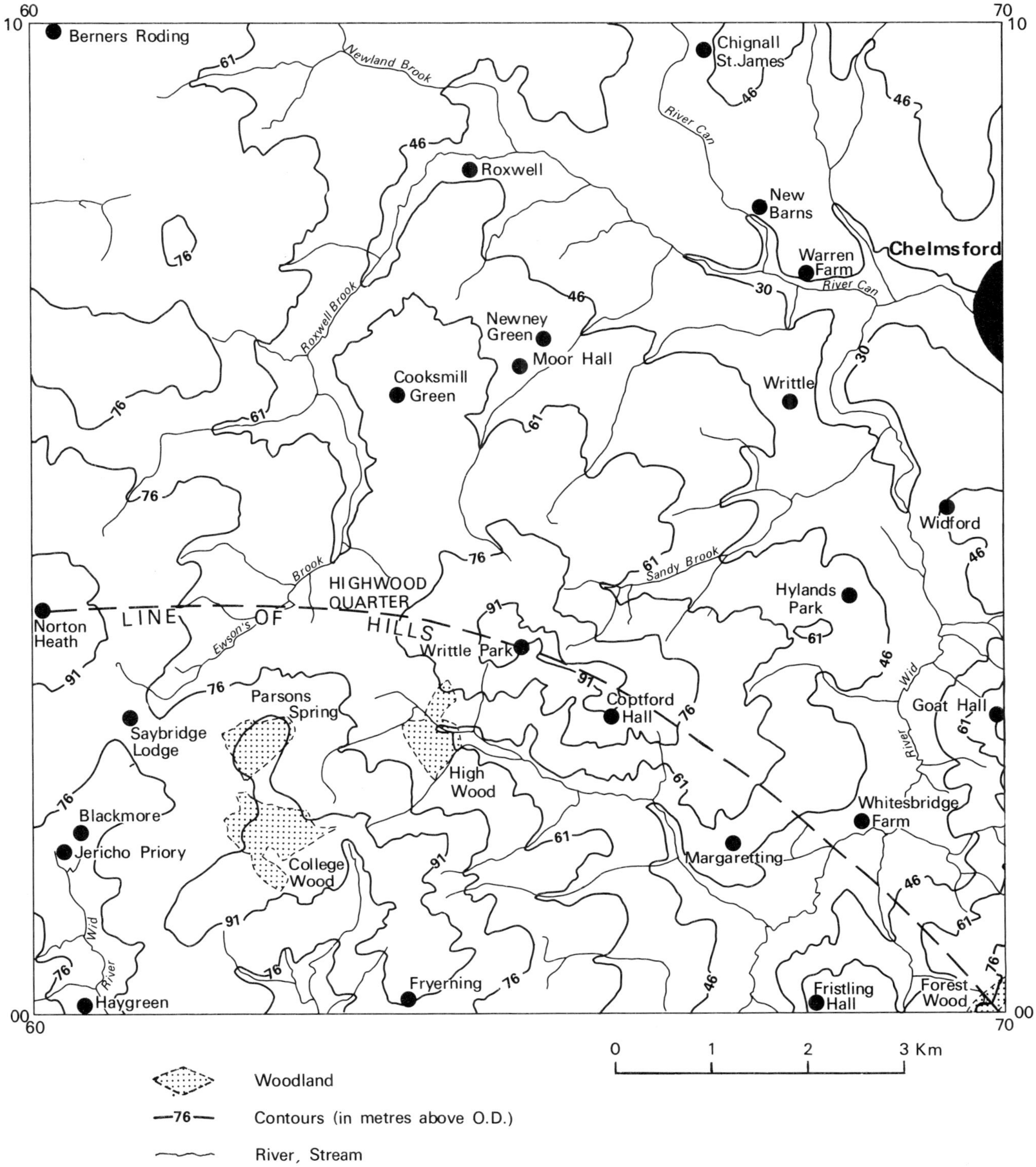

Figure 2 Sketch map of the topography of the country west of Chelmsford, showing the places mentioned in the text.

DESCRIPTION OF THE RESOURCE SHEET

GENERAL

The resource sheet covers an area of 100 km^2 to the west of Chelmsford, the county town of Essex (Figure 1). Chelmsford situated 48 km north-east of central London, is the centre for a large commuting population, and with Writtle, a small dormitory town to the west, covers an area of 7.5 km^2 on the eastern margin of the sheet. The remainder of the district is largely rural and the major industry is mixed farming, which includes some forestry on the higher ground towards the south. However, some light industries are located in the urban area of Chelmsford.

Essex has long been a source of aggregate and this sheet area, although only containing 17.6 km^2 of mineral-bearing ground, covers the western margin of a highly productive area centred to the east of Chelmsford.

TOPOGRAPHY

The area is dominated to the south by a line of hills, the remnants of a dissected upland, which extends from Norton Heath [602 043] in the west to Stock [TQ 688 986] in the south-east (Figures 1 and 2). These hills, which rise to 98 m above Ordnance Datum, are broken by two gaps. In the west at Saybridge Lodge [610 030], between Norton Heath and Blackmore [605 018], Ewson's Brook cuts a broad shallow depression trending north-eastwards, and in the east the River Wid, flowing in a narrow deep valley, cuts through these

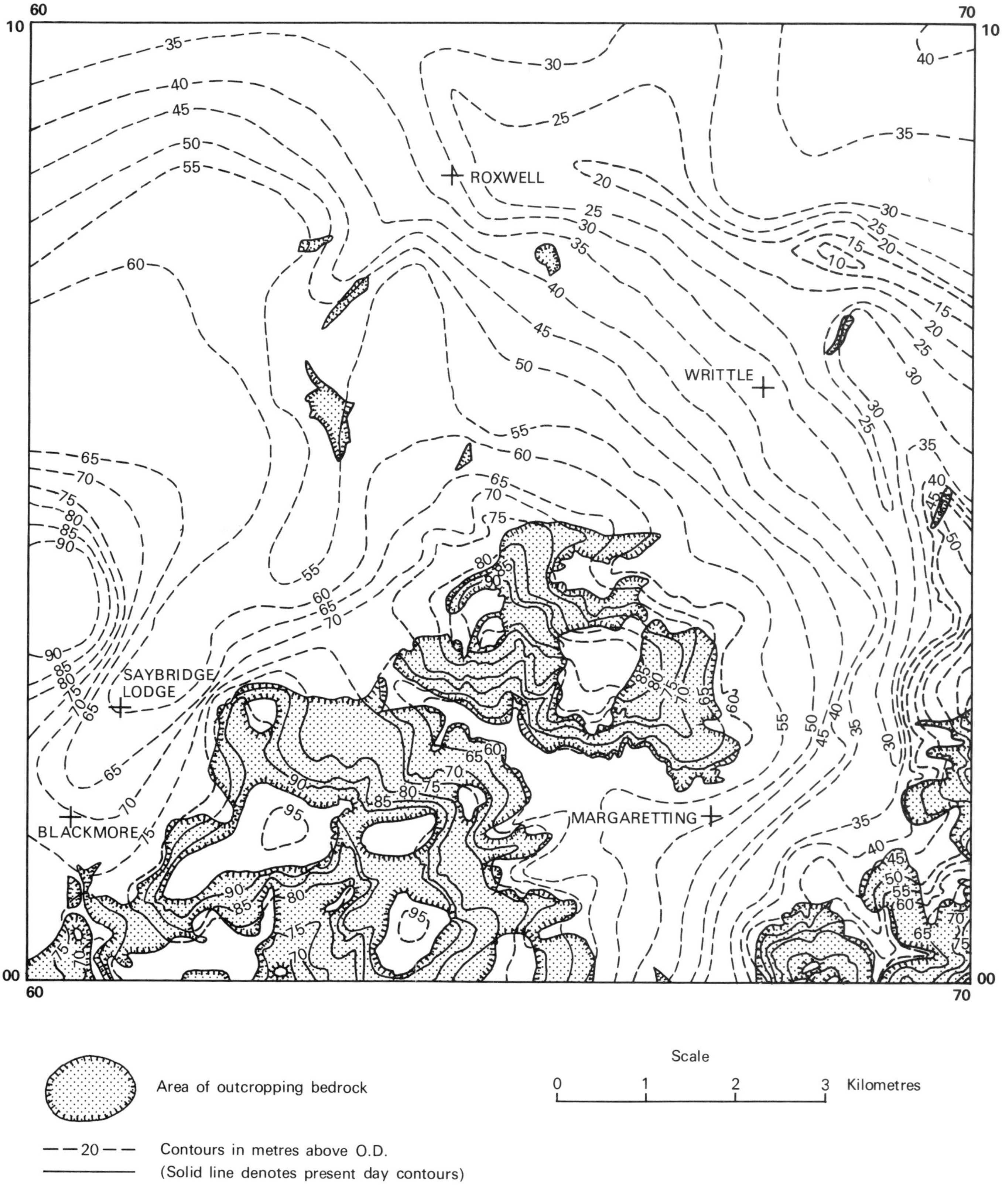

Figure 3 Contour map of the bedrock surface.

hills south-east of Margaretting [671 017]. The River Wid rises at Jericho Priory [603 016], Blackmore, and flows south leaving the sheet area at Hay Green [650 000] (Figure 1). It then flows south-eastwards south of Mountnessing [TQ 632 976] before turning north-east and re-entering the sheet area near Fristling Hall [680 000]. Thence it flows northwards from Whitesbridge Farm [684 019] to its confluence with the River Can east of Warren Farm [680 074] (Figure 2).

North of these hills the ground falls gently towards the north-east and is dissected by numerous streams. Ewson's and Newland Brooks are tributaries of the Roxwell Brook which flows north from near Ewson's Bridge [633 050] to Roxwell [645 084] and then east to join the River Can.

The River Can flows south-east from Chignall St James [670 096] turning to the east at its confluence [675 074] with Roxwell Brook. After being joined by the River Wid it flows east through Chelmsford to its confluence with the Chelmer.

In the north-west, near Spains Hall [603 064] the ground rises to 80 m above Ordnance Datum at the

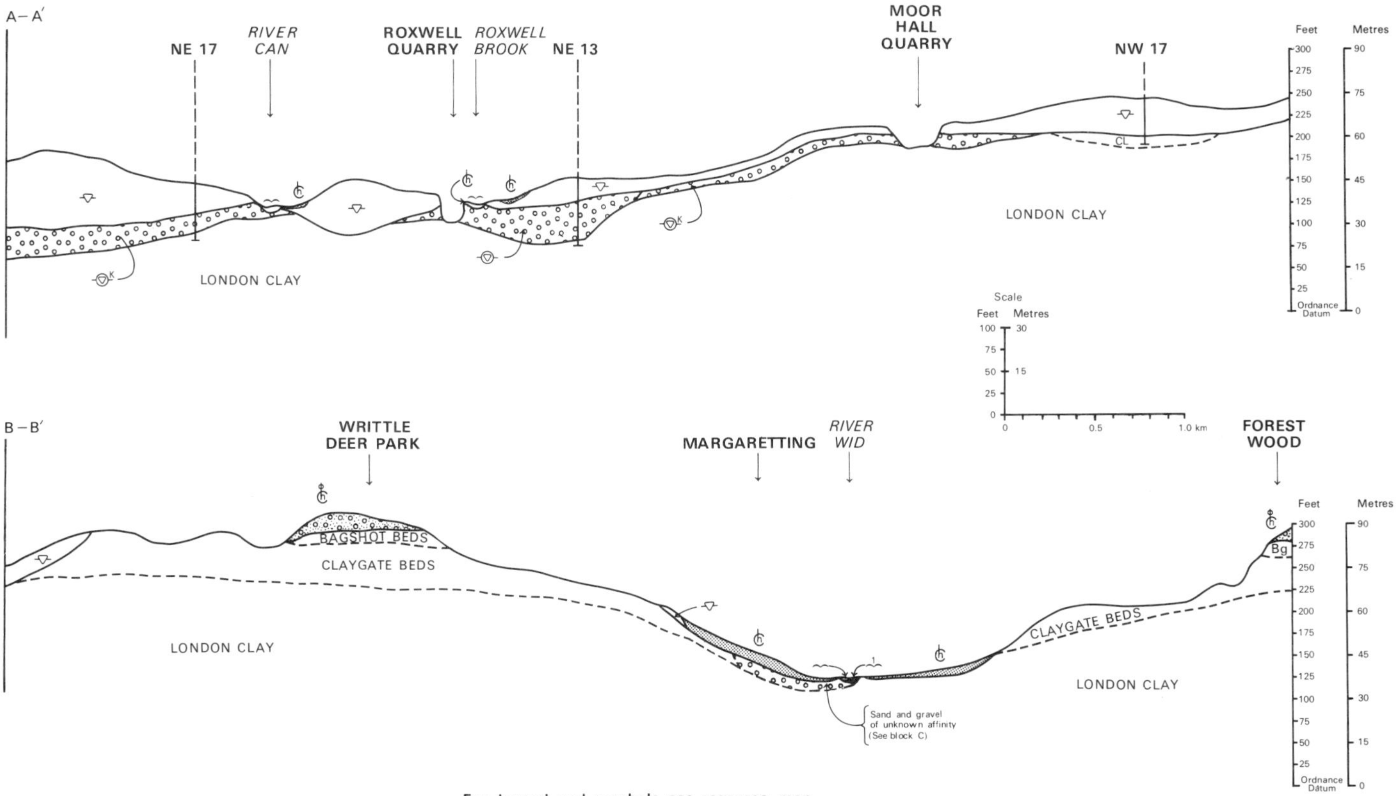

Figure 4 Sketch section showing the general relationships of the drift and bedrocks.

southern margin of the Boulder Clay sheet that extends over most of the northern part of this district.

GEOLOGY

This district may be divided into two areas of contrasting geological characteristics, lying respectively north and south of the line of hills mentioned above (Figure 2).

The northern area is a boulder-clay plain with infrequent and small patches of Kesgrave Sands and Gravels and small inliers of London Clay; to the south large outcrops of London Clay, Claygate Beds and Bagshot Beds are capped on the higher ground by patches of Older Head and in the valleys by Head (Figures 2 and 4). This general pattern is modified in the gap at Saybridge Lodge [610 030] where Boulder Clay and Glacial Sand and Gravel predominate, and in the valleys of the Rivers Wid, Can and Roxwell Brook where Alluvium and Head deposits occur.

The Upper Chalk, Thanet Beds and Woolwich and Reading Beds are encountered only in deep boreholes beneath a considerable thickness of London Clay and will not be discussed further in this report.

A list of the formations that crop out in the area of the resource sheet is given in Table 1.

SOLID FORMATIONS

London Clay

The London Clay crops out as small inliers surrounded by the overlying Boulder Clay in the valley of Roxwell Brook and in the vicinity of Newney Green [651 069]; it forms more extensive outcrops around Blackmore and High Woods [63 02].

The London Clay consists predominantly of silty clay or clayey silt but at both the top and bottom of the formation it includes a fine sand. Characteristically at the top of the formation on alternation of sand and clay lithologies occurs, with the sand concentrated within fine laminae and thin seams separated by silty clay.

The boundary between the London Clay and the overlying Claygate Beds is gradational, but during mapping the lowest significant sandy horizon recognis-

Table 1 List of the formations that crop out in the area of the resource sheet

Recent and **Pleistocene**	
Peat	Dark brown or black vegetable matter with varying amounts of silt and sand
Alluvium	Clayey silt and silty fine sand with some thin gravel seams, yellowish brown
First Terrace	Gravel and sandy gravel, yellowish brown
Head	Silty sandy clay with some flint pebbles, medium brown
Glacial Silts	Laminated silt and fine sand with some gastropod debris and peat seams, yellowish grey
Boulder Clay	Silty clay with flint and chalk pebbles and some fine sand, dark bluish grey, weathering to ochre brown
Glacial Sand and Gravel	Gravel with some thin pebbly silty boulder clay seams
Kesgrave Sands and Gravels	Sandy gravel with some thin silty clay seams, yellowish brown and orange-brown
Older Head	'Very clayey' well rounded flint gravel, yellowish brown and orange-brown
Eocene	
Bagshot Beds	Fine and medium sand with subordinate thin clay seams, yellowish brown and orange-brown
Claygate Beds	Fine sandy clay and silt, laminated, pale yellowish brown
London Clay	Dark bluish grey silty clay with some fine sand

able in the field was used as the boundary. Throughout the London Clay, pockets or lenses of drab olive-green glauconitic fine and medium sand may occur.

The colour of the London Clay varies from dark olive-green to dark bluish grey when fresh, but on weathering the clay is usually brown with orange-brown mottling. Pyrite occurs throughout the deposit both as disseminated particles and as nodules. The reaction between the weathering products of pyrite and calcium carbonate present in the clay leads to the formation of scattered selenite (gypsum) crystals. Calcareous concretions (septarian nodules) occur at some levels within the clay; they are characteristically veined with calcite and often contain well preserved molluscan shells.

Claygate Beds

The passage beds which were deposited during the transitional period of gradual shallowing from the deeper-water conditions that prevailed during the deposition of the London Clay to the shallower-water conditions in which the Bagshot Beds were laid down, are called the Claygate Beds. They are a series of alternations of evenly-bedded and laminated sand, silt and clay. In the lower part of the sequence, near the junction with the London Clay, clay and silt seams predominate over thin partings of white sand, while in the upper part the sandy beds increase to the virtual exclusion of the clay. The Claygate Beds pass without a clear demarcation line into the overlying Bagshot Beds.

The sand is predominantly fine grained and yellow or pale yellow, whilst the clay and silt are white, pale grey or more commonly lilac grey.

Results from IGS stratigraphical boreholes at Stock, Hadleigh, Hockley and Westley Heights (Cratchley and others, 1975) have shown that three broad subdivisions of the Claygate Beds can be recognised: the lowest division is a micaceous glauconitic sand overlain by silty clay. The middle division is of finely laminated micaceous glauconitic fine-grained sand overlain by silty clay and clayey silt interlaminated with fine sand. The uppermost division consists of interbedded fine-grained sands and silts with subordinate clay lenses and wisps. These divisions cannot generally be recognised in IMAU boreholes since insufficient thicknesses of bedrock have been penetrated, but in borehole SE 83 [6558 0298] the middle and lower divisions are distinguishable.

Bagshot Beds

The Bagshot Beds are marine in origin and only their lower part is present in the area covered by this resource sheet. This consists of fine and medium slightly micaceous sands (about 13 m was proved in the IGS borehole at Stock), often laminated and cross-bedded with subordinate beds of lilac and light grey clay. The overlying clay and silt group, with pebble beds, is not seen in this area.

The sands range in colour from almost white to orange-brown but are generally pale brown and often contain black carbonaceous flecks. Subordinate beds of clay and silt are lilac or light grey in colour.

The base of the deposit lies generally at about 91 m above Ordnance Datum and the Bagshot Beds cap the highest ground around Writtle Park [649 036], Coptford Hall [659 030], College Wood [62 01], High Woods, Goat Hall [699 030] and Forest Wood [698 001].

Since in south-east Essex the top of the Bagshot Beds has been removed by post-Eocene erosion, the full original thickness is not known in this district.

DRIFT DEPOSITS

Older Head

This deposit is found in numerous small outcrops capping the high ground on the southern half of the sheet. The gravel fraction is predominantly (90 per cent) of rounded to well rounded flint set in a 'clayey' or 'very clayey' sand matrix. In some areas a pebbly sand is developed at the base, notably at Writtle Park and Parsons Spring [62 01], 1.5 km south-west of Highwood. A further variation from the normal lithology was proved in borehole SW 27 [6222 0149] and Exposure SW E1 [6234 0170] at College Wood, where the composition was highly variable, with much smaller amounts of well rounded flint and greater amounts of angular flint, quartzite, quartz and sandstone.

The origin and age of the Older Head deposit are not known with certainty but they are generally considered to be early Pleistocene in age. Prestwich (1890, p. 164) grouped these deposits with the Warley Gravels, found to the south of this sheet, and called them the Brentwood Group. He considered them to be 'little else than Bagshot Shingle, more or less disturbed, mixed with a few subangular flints, quartzite and quartz'.

The Older Head is probably a solifluxion deposit formed from the Bagshot Pebble beds mixed in varying proportions with high level gravels such as the Plateau Gravels of the Epping ridge, south-west of this sheet.

Kesgrave Sands and Gravels

This deposit, the principal occurrence of mineral on the sheet area, is limited in outcrop but has been proved to be more extensive under the boulder-clay cover. It occurs exclusively in the north where small outcrops are seen in the valleys of the River Can and Roxwell Brook. The gravels comprising the deposit have no limestone material and a high proportion (30 per cent) of quartz and quartzite. They were formerly thought to be outwash gravels, deposited in front of an advancing Pleistocene (Anglian) ice sheet. More recently, however, they have been considered to be periglacial (Beestonian) proto-Thames fluvial gravels, deposited in a braided-stream environment (Rose and Allen, 1977).

The Kesgrave Sands and Gravels thin towards the south where Boulder Clay rests directly on bedrock (Figure 4) as shown in boreholes NW 17 [6393 0519] and NE 16 [6516 0564].

Glacial Sand and Gravel

The Glacial Sand and Gravel, locally called 'Hussick', is more widely distributed, but often thinner than Kesgrave Sands and Gravels, being found both north and south of the hills capped by Bagshot Beds. It is a proximal outwash gravel closely associated with and intercalated within the Boulder Clay, and always rests on the older Kesgrave Sands and Gravels where the latter are present, for example, in borehole NW 15 [6303 0742]. The gravels found infilling the incised buried valley of the River Can/Roxwell Brook (Figure 3) are of this chalk-rich type, and have been recorded, for example, in boreholes NE 13 [6501 0814] and NE 24 [6835 0759].

Boulder Clay

Boulder Clay covers most of the northern half of this sheet and extends southwards in two lobes at Hylands Park [685 042] and in the col at Saybridge Lodge. A number of small patches are also found in the vicinity of Margaretting and Fryerning [638 002] in the south.

It consists of dark bluish grey silty sandy clay containing many pebbles of various sizes and rock-types; despite its name boulders are rare. The clay matrix is derived mainly from the pyrite-rich dark bluish grey and black Lias, Oxford, Ampthill, Kimmeridge and Gault clays, over which the ice moved on its journey south. The pebbles, constituting up to 40 per cent of the deposit, are predominantly of chalk with flint (both angular and well rounded) and some quartzite, quartz and black shale and a trace of fossil debris.

Weathering greatly affects the appearance of the Boulder Clay; generally the upper three or four metres have been leached of chalk and the pyritous clay matrix has been oxidised to produce a colour of brown or ochre. The base of this deposit, particularly where it rests directly on bedrock, becomes sandy and often very pebbly (but free of chalk), for example in borehole NW 12 [6183 0617].

Glacial Silts

Glacial Silts have been found in only two limited areas in this sheet area: in the north-east, half a kilometre east of New Barn [675 081], and in the south in the vicinity of Saybridge Lodge. In the north they crop out on the western flank of the valley of a tributary stream of the River Can and have been proved to the east beneath head deposits in borehole NE 24. They typically consist of laminated pale yellowish brown clayey silty fine sand and fine sandy silt with, high in the sequence, seams of dark brown peat and comminuted gastropod debris. Usually, a 'Clayey' gravel with disseminated carbonaceous material overlies the sand and silt, the organic matter indicating that the lake persisted into the subsequent warmer climatic period. In the south the Glacial Silts crop out around Saybridge Lodge and pass under Boulder Clay to the north-east; here they consist of silty chalky fine-grained sand, interlaminated with fine sandy silt, as proved in boreholes SW 16 [6099 0291] and SW 17 [6135 0319].

Head

This solifluxion deposit covers large areas in the vicinity of Writtle; it is also widespread in the valley of the River Wid, and in most of the other river valleys where it flanks the Alluvium. It is a heterogeneous deposit, whose lithology varies widely from silty clay through sandy pebbly silty clay to sandy gravel seams. These are thin and not considered potentially workable. Though up to 7.9 m of Head has been proved, for example, in borehole SE 89 [6981 0415], it is generally about two or three metres thick.

In the vicinity of Margaretting the Head masks gravel, probably of fluvial origin, which, for simplicity, has been classified with the Head on the map.

First Terrace

Only one small area of river terrace deposits, classified as First Terrace, has been mapped on the sheet area; it is south of Whitesbridge Farm [684 020] and is thought to be mineral, although no borehole information is available.

Alluvium

Alluvium, of silty sandy clay lithology, forms sinuous belts that closely follow the present-day streams. The Alluvium masks sandy gravel of fluvial origin in the valley of the River Wid.

Peat

This deposit occurs only in the valley of Roxwell Brook south-east of Patience Bridge [633 071]. It consists of decomposed vegetable matter mixed with silty sandy clay.

COMPOSITION OF THE SAND AND GRAVEL

Six formations contain potentially workable sand and gravel deposits: Older Head, Kesgrave Sands and Gravels, Glacial Sand and Gravel, Glacial Silts, First Terrace and the fluvial gravels associated with the River Wid.

For the purpose of this report, the Glacial Silts and River Wid deposits have been considered with the Glacial Sand and Gravel. For the small area of First Terrace, no grading or compositional data are available, and although the First Terrace sands and gravels are considered to be potentially workable they will not be considered further in this section.

Although the Claygate Beds and Bagshot Beds would qualify as mineral under a rigid interpretation of the IMAU criteria they are considered not to be potentially workable owing to the fineness of their sand, the included deleterious accessory minerals (particularly iron oxides) and their clayey nature.

Older Head

The Older Head consists of well rounded pebbles packed in a stiff matrix of sandy silty, often waxy, clay. At its top in most boreholes up to 0.8 m of loose pebbles in a subsoil have been found. It grades as a 'clayey' gravel; locally, however, as in borehole SW 26 [6238 0275], 'clayey' pebbly sands and 'clayey' sands, similar in lithology to the Claygate Beds, are found at the base. The mean grading for the deposit is fines 17 per cent, sand 40 per cent and gravel 43 per cent. The sand is medium with fine and some coarse grade (rarely exceeding 7 per cent) and is composed of angular flint with subrounded and rounded quartz. The gravel is coarse and fine with a trace of cobbles and consists, predominantly, of well rounded flint (67 per cent) with angular flint (25 per cent) and some subrounded fine-grade quartz and quartzite with a trace of sandstone, igneous and metamorphic rocks. In the coarse gravel grade, well rounded flint constitutes up to 90 per cent of the deposit and only trace amounts of other rock types occur. In contrast, the Older Head mapped at College Wood shows a greater percentage of quartz and quartzite and a preponderance of angular (50 per cent) over well rounded (24 per cent) flint. This composition may be the result of post-depositional solifluxion and admixture with glacial material.

Kesgrave Sands and Gravels

This formation contains the principal deposits of potentially workable mineral in the district: it has a mean grading of fines 6 per cent, sand 50 per cent and gravel 44 per cent, giving an overall classification of sandy gravel; only two boreholes proved mineral with more than 10 per cent fines.

The sand is predominantly of medium grade with some coarse and fine material and is composed of subrounded quartz with some angular flint. In the coarse sand grade, angular flint and quartz are of equal proportions.

The gravel fraction is fine with coarse material: cobble grade was found in only one borehole. The gravel is predominantly angular flint (50 per cent) with well rounded flint (24 per cent) and with some subrounded quartz, quartzite and subangular sandstone with a trace of igneous and metamorphic material. (For the usage of 'with' and 'and' in this context, see Appendix C.)

Glacial Sand and Gravel
The Glacial Sand and Gravel, which is present in blocks A and C, has a mean grading of fines 7 per cent, sand 45 per cent and gravel 48 per cent (including 1 per cent of cobble grade), giving an overall classification of gravel.

The sand is predominantly medium with coarse and some fine grades and is composed principally of subrounded quartz and angular flint with traces of ironstone, chalk and limestone.

The gravel fraction is fine with coarse and a trace of cobble grade, and is predominantly angular flint (53 per cent) with rounded flint (16 per cent) and subrounded to rounded chalk (13 per cent) with some rounded to subrounded quartz, quartzite and subangular sandstone and characteristically (in trace amounts) limestone, ironstone, phosphatic nodules, pyrite, red chalk and fossil debris.

In the valley of the River Wid, the Head and Alluvium cover sand and gravel deposits whose affinity is not known. Only two IMAU boreholes were drilled into these deposits and they proved sandy gravel and gravel respectively, giving a mean grading of fines 9 per cent, sand 47 per cent and gravel 44 per cent including 1 per cent of cobble grade. Borehole SE 88 [6842 0195] proved mineral with a composition akin to the Glacial Sand and Gravel, while in borehole SE 85 [6662 0086] material with a composition intermediate between that of Older Head and Glacial Sand and Gravel was proved, suggesting reworking and admixture of these deposits. The latter borehole also contains 10 per cent of well rounded 'jasperised' flints derived from the Warley Gravels to the south (Dines, 1925 p. 23).

The Glacial Silts, in the vicinity of Warren Farm, also contain some potentially workable very chalky 'clayey' gravel resembling the Glacial Sand and Gravel. They rest on lacustrine peats and silts and Boulder Clay.

THE MAP

The sand and gravel resource map is folded into the pocket at the end of this report. The base map is the Ordnance Survey 1 : 25 000 Outline Edition in grey, on which the topography is shown by contours in green, the geological data in black and the mineral resource information in shades of red.

Geological data The geological boundary lines, symbols, etc., shown are taken from the geological map of this area recently surveyed at the scale of 1 : 10 560. This information was obtained by detailed application of field mapping techniques by the field staff of the Institute's East Anglia and South-Eastern England Unit. Borehole data, which include the stratigraphic relations, thicknesses and mean particle size distribution of the sand and gravel samples collected during the assessment, are also shown.

The geological boundaries are regarded as the best interpretation of the information available at the time of the survey. However, it is inevitable, particularly with deposits (such as those represented in this area) which change rapidly vertically and laterally, that local irregularities or discrepancies will be revealed by some boreholes (as, for example, at borehole SE 84). These are taken into account in the assessment of the resources (see Appendix B).

Mineral resource information The mineral-bearing ground is subdivided into resource blocks (see Appendix A). Within a resource block the mineral is subdivided into areas where it is exposed, that is where overburden averages less than 1 m in thickness, and areas where it is present in continuous spreads beneath overburden. The 'almost continuous' or 'discontinuous' mineral categories have not been recognised in this area.

Areas where bedrock crops out, where boreholes indicate absence of sand and gravel beneath cover and where sand and gravel beneath cover is interpreted to be not potentially workable, are uncoloured on the map; where appropriate the relevant criterion is noted. In such cases it has been assumed that mineral is absent except in infrequent and relatively minor patches which can neither be outlined nor assessed quantitatively in the context of this survey. Areas of unassessed sand and gravel, for example in built-up areas, are indicated by a red stipple.

The area of the sand and gravel is measured, where possible, from the mapped geological boundary lines. The whole of this area is considered as mineral, although it may include small areas where sand and gravel is not present or is not potentially workable. Inferred boundaries have been inserted where sand and gravel beneath cover is interpreted to be not potentially workable or absent. Such boundaries (for which a distinctive symbol is used) are drawn primarily for the purpose of volume estimation. The symbol is intended to convey an approximate location within a likely zone of occurrence rather than to represent the breadth of the zone, its size being limited only by cartographic considerations. For the purpose of measuring areas the centre-line of the symbol is used.

Worked areas and made ground The approximate extent of known sand and gravel workings to July 1979 are shown on the map; active and disused workings are indicated, together with areas which have been returned to agricultural use and areas partly backfilled with waste from the sand and gravel industry. Workings which have been filled with mixed waste or refuse are shown as 'made ground'.

RESULTS

The statistical results are summarised in Table 2. Fuller grading particulars are shown in Figure 5 and Tables 3, 4 and 5.

Accuracy of results For the three resource blocks A, B and C the accuracy of the results at the 95 per cent probability level ranges from 32 to 48 per cent (that is, it is probable that 19 times out of 20 the true volumes lie within the given limits of the mean). However, the true values are more likely to be nearer the figures estimated than the limits. Moreover, it is probable that in each block approximately the same percentage limits would apply for the estimate of volume of a very much smaller parcel of ground (say 200 acres) containing similar sand and gravel deposits, if the results from the same number

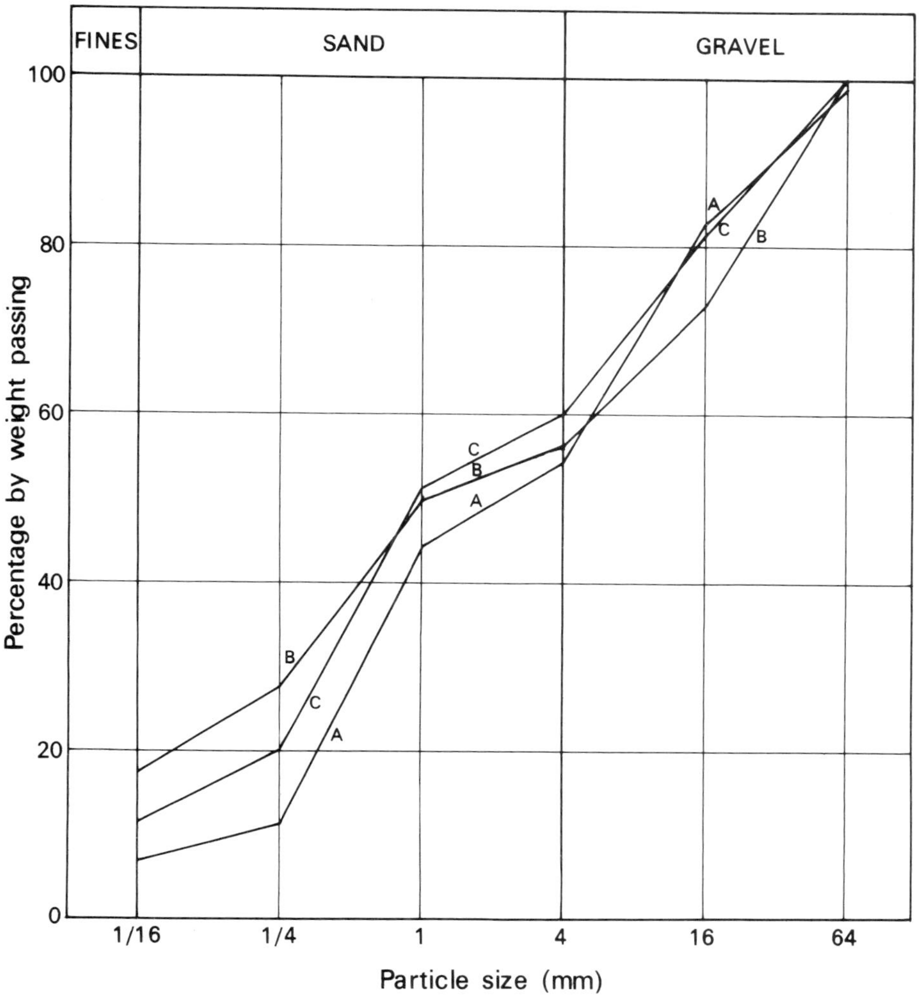

Figure 5 Particle size distribution for the assessed thickness of mineral in resource blocks A, B and C.

Resource Block	Percentage by weight passing					
	1/16mm	1/4mm	1mm	4mm	16mm	64mm
A	6	11	44	54	81	100
B	17	28	50	56	73	100
C	11	20	51	60	82	99

of sample points (as provided by, say, ten boreholes) were used in the calculation. Thus, if closer limits are needed for the quotation of the reserves in part of a block, it can be expected that data from more than ten sample points will be required, even if the area is quite small. This point can be illustrated by considering the whole of the potentially workable sand and gravel on this sheet. The total volume (66.9 million m^3) can be estimated to limits of ±23 per cent at the 95 per cent probability level, by a calculation based on the data from eighty-eight sample points spread across the three resource blocks. However, it must be emphasised that the quoted volume of sand and gravel has no simple relationship with the amount that could be extracted in practice, as no allowance has been made in the calculations for any restraints (such as existing buildings and roads) on the use of land for mineral working.

Table 2 The sand and gravel resources of sheet TL 60

Block	Area		Mean thickness		Volume of mineral			Mean grading percentages					
	Block	Mineral	Over-burden	Mineral		Limits at the 95% confidence level		Fines	Sand			Gravel	
	km^2	km^2	m	m	$10^6\ m^3$	±%	±Volume $10^6\ m^3$	$-\frac{1}{16}$ mm	$+\frac{1}{16}-\frac{1}{4}$ mm	$+\frac{1}{4}-1$ mm	+1–4 mm	+4–16 mm	+16 mm
A	40.1	10.5	5.7	4.9	51.5	33	17	6	5	33	10	27	19
B	29.6	2.1	0.4	3.0	6.3	32	2	17	11	22	6	17	27
C	23.8	5.0	2.3	2.7	13.5	48	7	11	9	31	9	22	18
Total (A, B, C)	92.5	17.6	3.5	3.8	66.9	23	15	9	7	30	9	24	21

NOTES ON THE RESOURCE BLOCKS

The area of sheet TL 60 (excluding the urban areas of Writtle and Chelmsford) has been divided into three resource blocks: Block A in the north contains the principal deposits of Kesgrave Sands and Gravels, and some Glacial Sand and gravel. Block B the Older Head and Block C the remaining mineral deposits proved during the survey.

Block A

This block covers 40.1 km^2, of which 10.5 km^2 is mineral-bearing; it extends over the northern half of the sheet with the exception of the urban areas of Writtle and Chelmsford, and a small area of Block C immediately to the west of Chelmsford.

Much of the sand and gravel within the rural area is concealed beneath Boulder Clay with only limited outcrops in the valleys of the Roxwell Brook and River Can. The overburden within the mineral area ranges in thickness from 1.1 m in borehole NW 19 [6490 0648] to 10.4 m in borehole NW 11 [6181 0731] (Table 3) and has a mean thickness of 5.7 m. The Boulder Clay becomes excessively thick outside the mineral area, and the underlying Kesgrave Sands and Gravels and Glacial Sand and Gravel thin out against the hills in the south-west and south (Figure 4).

Table 3 Data from IMAU boreholes: Block A

Borehole No.	Recorded thickness		Mean grading percentage					
	Mineral	Overburden	Fines	Fine sand	Medium sand	Coarse sand	Fine gravel	Coarse gravel
	m	m	$-\frac{1}{16}$ mm	$+\frac{1}{16}-\frac{1}{4}$ mm	$+\frac{1}{4}-1$ mm	+1–4 mm	+4–16 mm	+16 mm
NW 8	absent	–	–	–	–	–	–	–
NW 9	absent	–	–	–	–	–	–	–
NW 10	6.6	5.4	14	5	45	6	19	11
NW 11	3.8	10.4	1	1	20	17	37	24
NW 12	absent	–	–	–	–	–	–	–
NW 13	1.6*	12.9	1	3	23	11	46	16
NW 14	2.8	9.1	1	3	31	8	31	26
NW 15	3.2†	4.5	2	1	58	14	18	7
NW 16	4.7	6.6	6	3	30	9	26	26
NW 17	absent	–	–	–	–	–	–	–
NW 18	2.5	2.1	4	3	17	15	37	24
NW 19	4.7	1.1	13	13	32	7	20	15
NE 12	absent	–	–	–	–	–	–	–
NE 13	14.0‡	1.4	6	4	26	14	29	21
NE 14	3.2	5.7	3	3	18	14	36	26
NE 15	absent	–	–	–	–	–	–	–
NE 16	absent	–	–	–	–	–	–	–
NE 17	3.7	9.4	3	5	33	9	28	22
NE 18	absent	–	–	–	–	–	–	–
NE 19	0.7*	1.3	4	3	30	8	31	24
NE 20	absent	–	–	–	–	–	–	–
NE 21	5.9	5.4	2	4	38	10	29	17
NE 22	absent	–	–	–	–	–	–	–
NE 23	7.1	1.9	6	6	49	6	19	14

* Sand and gravel that does not satisfy the criteria of 'mineral'.
† Excluding 2.2-m waste parting.
‡ Excluding 4.3-m waste parting.

Table 4 Data from IMAU boreholes: Block B

Borehole No.	Recorded thickness		Mean grading percentage					
	Mineral	Overburden	Fines	Fine sand	Medium sand	Coarse sand	Fine gravel	Coarse gravel
	m	m	$-\frac{1}{16}$ mm	$+\frac{1}{16}-\frac{1}{4}$ mm	$+\frac{1}{4}-1$ mm	+1–4 mm	+4–16 mm	+16 mm
SW 23	3.1	0.2	12	4	25	6	15	38
SW 26	3.9*	1.0	20	8	39	5	11	17
SW 27	4.7	0.1	10	3	20	10	31	26
SW 28	3.5	0.1	14	8	17	5	16	40
SW 29	4.6	0.1	24	15	26	4	13	18
SW E1	2.3+	0.5	16	3	7	7	25	42
SE 83	2.4	0.1	22	18	20	3	12	25
SE 84	absent	–	–	–	–	–	–	–
SE 90	1.2	0.3	29	63	3	2	1	2
SE 91	absent	–	–	–	–	–	–	–

* Excluding 2.3-m waste parting.

Table 5 Data from IMAU boreholes: Block C

Borehole No.	Recorded thickness		Mean grading percentage					
	Mineral	Overburden	Fines	Fine sand	Medium sand	Coarse sand	Fine gravel	Coarse gravel
	m	m	$-\frac{1}{16}$ mm	$+\frac{1}{16}-\frac{1}{4}$ mm	$+\frac{1}{4}-1$ mm	+1–4 mm	+2–16 mm	+16 mm
NE 24	6.1*	1.4	14	14	32	8	18	14
NE 25	absent	–	–	–	–	–	–	–
SW 24	1.2	2.0	2	3	22	10	29	34
SW 25	absent	–	–	–	–	–	–	–
SE 85	3.3	0.1	9	5	32	8	25	21
SE 86	absent	–	–	–	–	–	–	–
SE 87	absent	–	–	–	–	–	–	–
SE 88	1.2	2.7	8	6	31	17	28	10
SE 89	absent	–	–	–	–	–	–	–

* Excluding 5.6-m waste parting.

The mineral ranges in thickness from 2.5 m in borehole NW 18 [6428 0738] to a total of 14.0 m in borehole NE 13 [6501 0814] and has a mean thickness of 4.9 m. The great thickness of sand and gravel in borehole NE 13 represents an infill of chalky Glacial Sand and Gravel in the deeply cut buried channel of the 'proto' Can (Figure 4).

The mean grading for the block is fines 6 per cent, sand 48 per cent and gravel (including a trace of cobble grade) 46 per cent (Table 2 and Figure 5). The estimated total volume of mineral is 51.5 ± 17 million m^3.

Block B

This block, which is divided into two areas by the valley of the River Wid, covers 29.6 km^2 of ground around Highwood and to the south-east of Margaretting; of this area, only 2.1 km^2 is mineral-bearing.

Much of the mineral is exposed with only small areas concealed beneath Head around College Wood, Norton Heath and Fryerning. The overburden ranges in thickness from 0.1 m, for example in borehole SE 83 [6558 0298], to 1.0 m in borehole SW 26 [6238 0275] and has a mean thickness of 0.4 m (Table 4).

Borehole SE 91 [6997 0005] in the extreme south-east of the sheet was drilled into the edge of an Older Head deposit. It proved no potentially workable sand and gravel.

The mineral has a mean thickness of 3.0 m and ranges in thickness from 1.2 m in borehole SE 90 [6983 0282] to 4.7 m in borehole SW 27 [6222 0149]. The block contains an estimated total volume of 6.3 ± 2 million m^3 of mineral: its mean grading is fines 17 per cent, sand 39 per cent and gravel (including a trace of cobble grade) 44 per cent (Table 2).

Block C

This block, containing all the remaining glacial, together some fluvial mineral deposits, covers an area of 23.8 km^2, of which 5.0 km^2 is mineral-bearing. The overburden, consisting of Alluvium, Head, Boulder Clay and Glacial Silts, ranges in thickness from 0.1 m in borehole SE 85 [6662 0086] to 5.0 m in borehole SE 50 [6842 0251] and has a mean thickness of 2.3 m (Table 5). The mean recorded thickness of the mineral is 2.7 m and ranges from 6.1 m in borehole NE 24 [6835 0759] to 1.2 m in, for example, borehole SE 88 [6835 0195] (Table 2).

The total volume of mineral in the block is estimated at 13.5±7 million m^3: its mean grading is fines 11 per cent, sand 49 per cent and gravel 40 per cent, including 1 per cent of cobble grade.

NOTES ON THE SAND AND GRAVEL WORKINGS OF THE AREA

Sand and gravel has been worked from both the Kesgrave Sands and Gravels and Older Head deposits of the district (see table 6). To date workings cover an area of 74 hectares and an estimated 3.5 million m^3 has been removed. Of this total only about 0.3 million m^3 has been won from the Older Head. The Older Head workings have been abandoned since the turn of the century. They were worked for hoggin, on a local basis, to provide hardcore for tracks and roads. These hand-dug pits have not been reclaimed and they are now heavily overgrown or flooded.

The Kesgrave Sands and Gravels pit at Chignall St James and the Glacial Sand and Gravel Pits at Roxwell are being or have been reclaimed and the working face at Chignall is now north of the sheet margin.

Table 6 List of active and disused workings

Location and Grid Reference	Deposit worked
Active workings	
Chignall St James [662 100]	Kesgrave Sands and Gravels
Moor Hall 1 [656 065]	Kesgrave Sands and Gravels
Moor Hall 2 [648 065]	Kesgrave Sands and Gravels
Main disused workings	
Roxwell 1 [657 085]	Glacial Sand and Gravel
Roxwell 2 [664 081]	Glacial Sand and Gravel
Norton Heath [603 043]	Older Head
Parsons Spring [624 027]	Older Head
College Wood [623 015]	Older Head
Writtle Deer Park [654 029]	Older Head

APPENDIX A

FIELD AND LABORATORY PROCEDURES

Trial and error during initial studies of the complex and variable glacial deposits of East Anglia and Essex showed that an absolute minimum of five sample points evenly distributed across the sand and gravel are needed to provide a worthwhile statistical assessment, but that, where possible, there should be not less than ten. Sample points are any points for which adequate information exists about the nature and thickness of the deposit and may include boreholes other than those drilled during the survey and exposures. In particular, the cooperation of sand and gravel operators ensures that boreholes are not drilled where reliable information is already available; although this may be used in the calculations, it is held confidentially by the Institute and cannot be disclosed.

The mineral shown on each 1:25 000 sheet is divided into resource blocks. The arbitrary size selected, 10 km^2, is a compromise to meet the aims of the survey by providing sufficient sample points in each block. As far as possible the block boundaries are determined by geological boundaries so that, for example, glacial and river terrace gravels are separated. Otherwise division is by arbitrary lines, which may bear no relationship to the geology. The blocks are drawn provisionally before drilling begins.

A reconnaissance of the ground is carried out to record any exposures and inquiries are made to ascertain what borehole information is available. Borehole sites are then selected to provide an even pattern of sample points at a density of approximately one per square kilometre. However, because broad trends are independently overlain by smaller scale characteristically random variations, it is unnecessary to adhere to a square grid pattern. Thus such factors as ease of access and the need to minimise disturbance to land and the public are taken into account in siting the holes; at the same time it is necessary to guard against the possibility that ease of access (that is, the positions of roads and farms) may reflect particular geological conditions, which may bias the drilling results.

The drilling machine employed should be capable of providing a continuous sample representative of all unconsolidated deposits, so that the in-situ grading can be determined, if necessary, to a depth of 30 m at a diameter of about 200 mm, beneath different types of overburden. It should be reliable, quiet, mobile and relatively small (so that it can be moved to sites of difficult access). Shell and auger rigs have proved to be almost ideal.

The rigs are modified to enable deposits above the water table to be drilled 'dry', instead of with water added to facilitate the drilling, to minimise the amount of material drawn in from outside the limits of the hole. The samples thus obtained are representative of the in-situ grading, and satisfy one of the most important aims of the survey. Below the water table the rigs are used conventionally, although this may result in the loss of some of the fines fraction and the pumping action of the bailer tends to draw unwanted material into the hole from the sides or the bottom.

A continuous series of bulk samples is taken throughout the sand and gravel. Ideally samples are composed exclusively of the whole of the material encountered in the borehole between stated depths. However, care is taken to discard, as far as possible, material which has caved or has been pumped from the bottom of the hole. A new sample is commenced whenever there is an appreciable lithological change within the sand and gravel, or at every 1 m depth. The samples, each weighing between 25 and 45 kg, are despatched in heavy duty polythene bags to a laboratory for grading. The grading procedure is based on British Standard 1377 (1967). Random checks on the accuracy of the grading are made in the Institute's laboratories.

All data, including mean grading analysis figures calculated for the total thickness of the mineral, are entered on standard record sheets, abbreviated copies of which are reproduced in Appendix F.

Detailed records may be consulted at the appropriate offices of the Institute, upon application to the Head, Industrial Minerals Assessment Unit.

APPENDIX B

STATISTICAL PROCEDURE

Statistical assessment

1 A statistical assessment is made of an area of mineral greater than 2 km^2, if there is a minimum of five evenly spaced boreholes in the resource block (for smaller areas see paragraph 12 below).

2 The simple methods used in the calculations are consistent with the amount of data provided by the survey. Conventional symmetrical confidence limits are calculated for the 95 per cent probability level, that is, there is a 5 per cent or one in twenty chance of a result falling outside the stated limits.

3 The volume estimate (V) for the mineral in a given block is the product of the two variables, the sampled areas (A) and the mean thickness ($\bar{l}_m$) calculated from the individual thicknesses at the sample points. The standard deviations for these variables are related such that

$$S_V = \sqrt{(S_A{}^2 + S_{\bar{l}_m}{}^2)} \quad . \qquad [1]$$

4 The above relationship may be transposed such that

$$S_V = S_{\bar{l}_m}\sqrt{(1 + S_A{}^2/S_{\bar{l}_m}{}^2)} \quad . \qquad [2]$$

From this it can be seen that as $S_A{}^2/S_{\bar{l}_m}{}^2$ tends to 0, S_V tends to $S_{\bar{l}_m}$.

If, therefore, the standard deviation for area is small with respect to that for mean thickness, the standard deviation for volume approximates to that for mean thickness.

5 Given that the number of approximately evenly spaced sample points in the sampled area is n with mineral thickness measurements $l_{m_1}, l_{m_2}, \ldots l_{m_n}$, then the best estimate of mean thickness, l_m, is given by

$$\Sigma(l_{m_1} + l_{m_2} \ldots l_{m_n})/n \quad .$$

For groups of closely spaced boreholes a discretionary weighting factor may be applied to avoid bias (see note on weighting below). The standard deviation for mean thickness $S_{\bar{l}_m}$, expressed as a proportion of the mean thickness, is given by

$$S_{\bar{l}_m} = (1/\bar{l}_m)\sqrt{[\Sigma(l_m - \bar{l}_m)^2/(n-1)]}$$

where l_m is any value in the series l_{m_1} to l_{m_n}.

6 The sampled area in each resource block is coloured pink on the map. Wherever possible, calculations relate to the mineral within mapped geological boundaries (which may not necessarily correspond to the limits of deposit). Where the area is not defined by a mapped boundary, that is, where the boundary is inferred, a distinctive symbol is used. Experience suggests that the errors in determining area are small relative to those in thickness. The relationship $S_A/S_{\bar{l}_m} \leq \frac{1}{3}$ is assumed in all cases. It follows from equation [2] that

$$S_{\bar{l}_m} \leq S_V \leq 1.05\, S_{\bar{l}_m} \quad . \qquad [3]$$

7 The limits on the estimate of mean thickness of mineral, $L_{\bar{l}_m}$, may be expressed in absolute units $\pm(t/\sqrt{n}) \times S_{\bar{l}_m}$ or as a percentage $\pm(t/\sqrt{n}) \times S_{\bar{l}_m} \times (100/\bar{l}_m)$ per cent, where t is Student's t at the 95 per cent probability level for $(n-1)$ degrees of freedom, evaluated by reference to statistical tables. (In applying Student's t it is assumed that the measurements are distributed normally.)

Block calculation 1:25 000 Block Fictitious

Area
Block: 11.08 km^2
Mineral: 8.32 km^2

Mean thickness
Overburden: 2.5 m
Mineral: 6.5 m

Volume
Overburden: 21 million m^3
Mineral: 54 million m^3

Confidence limits of the estimate of mineral volume at the 95 per cent probability level: ±20 per cent
That is, the volume of mineral (with 95 per cent probability): 54 ± 11 million m^3

Thickness estimate measurements in metres
l_o = overburden thickness l_m = mineral thickness

Sample point	Weighting w	Overburden l_o	Overburden wl_o	Mineral l_m	Mineral wl_m	Remarks
SE 14	1	1.5	1.5	9.4	9.4	IMAU boreholes
SE 18	1	3.3	3.3	5.8	5.8	
SE 20	1	nil	–	6.9	6.9	
SE 22	1	0.7	0.7	6.4	6.4	
SE 23	1	6.2	6.2	4.1	4.1	
SE 24	1	4.3	4.3	6.4	6.4	
SE 17	$\frac{1}{2}$	1.2	1.6	9.8	7.2	Hydrogeology Unit record
123/45	$\frac{1}{2}$	2.0		4.6		
1	$\frac{1}{4}$	2.7	2.6	7.3	5.8	Close group of four boreholes (commercial)
2	$\frac{1}{4}$	4.5		3.2		
3	$\frac{1}{4}$	0.4		6.8		
4	$\frac{1}{4}$	2.8		5.9		
Totals	$\Sigma w = 8$	$\Sigma wl_o = 20.2$		$\Sigma wl_m = 52.0$		
Means		$w\bar{l}_o = 2.5$		$w\bar{l}_m = 6.5$		

Calculation of confidence limits

wl_m	$\lvert(wl_m - w\bar{l}_m)\rvert$	$(wl_m - w\bar{l}_m)^2$
9.4	2.9	8.41
5.8	0.7	0.49
6.9	0.4	0.16
6.4	0.1	0.01
4.1	2.4	5.76
6.4	0.1	0.01
7.2	0.7	0.49
5.8	0.7	0.49

$\Sigma(wl_m - w\bar{l}_m)^2 = 15.82$
$n = 8$
$t = 2.365$

L_V is calculated as

$t/w\bar{L}_m$

$$1.05(t/w\bar{l}_m)\sqrt{(\Sigma(wl_m - w\bar{l}_m)^2/n(n-1)]} \times 100$$
$$= 1.05 \times (2.365/6.5)\sqrt{[15.82/(8 \times 7)]} \times 100$$
$$= 20.3$$
$$\simeq 20 \text{ per cent}$$

Figure 6 Example of resource block assessment: calculation and results.

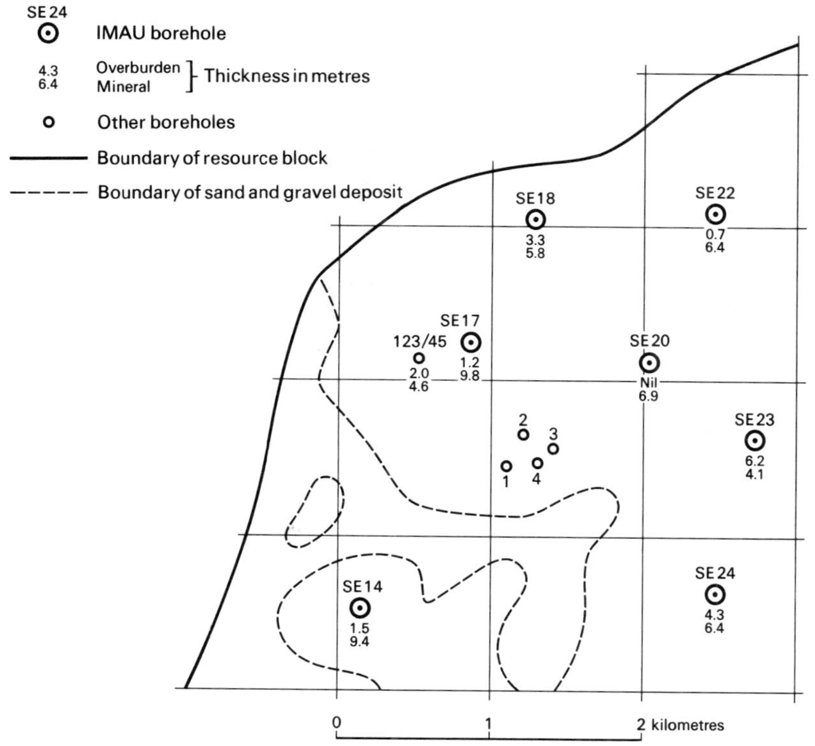

Figure 7 Example of resource block assessment: map of a fictitious block.

8 Values of t at the 95 per cent probability level for values of n up to 20 are as follows:

n	t	n	t
1	infinity	11	2.228
2	12.706	12	2.201
3	4.303	13	2.179
4	3.182	14	2.160
5	2.776	15	2.145
6	2.571	16	2.131
7	2.447	17	2.120
8	2.365	18	2.110
9	2.306	19	2.101
10	2.262	20	2.093

(from table 12, Biometrika Tables for Statisticians, Volume 1, Second Edition, Cambridge University Press, 1962). When n is greater than 20, 1.96 is used (the value of t when n is infinity).

9 In calculating confidence limits for volume, L_V, the following inequality corresponding to equation [3] is applied: $L_{\bar{l}_m} \leqslant L_V \leqslant 1.05\ L_{\bar{l}_m}$.

10 In summary, for values of n between 5 and 20, L_V is calculated as

$$[(1.05 \times t)/\bar{l}_m] \times [\sqrt{\Sigma(l_m - \bar{l}_m)^2/n(n-1)}] \times 100$$

per cent, and when n is greater than 20, as

$$[(1.05 \times 1.96)/\bar{l}_m] \times [\sqrt{\Sigma(l_m - \bar{l}_m)^2/n(n-1)}] \times 100$$

per cent.

11 The application of this procedure to a fictitious area is illustrated in Figures 6 and 7.

Inferred assessment

12 If the sampled area of mineral in a resource block is between 0.25 km^2 and 2 km^2 an assessment is inferred, based on geological and topographical information usually supported by the data from one or two boreholes. The volume of mineral is calculated as the product of the area, measured from field data, and the estimated thickness. Confidence limits are not calculated.

13 In some cases a resource block may include an area left uncoloured on the map, within which mineral (as defined) is interpreted to be generally absent. If there is reason to believe that some mineral may be present, an inferred assessment may be made.

14 No assessment is attempted for an isolated area of mineral less than 0.25 km^2.

15 *Note on weighting* The thickness of a deposit at any point may be governed solely by the position of the point in relation to a broad trend. However, most sand and gravel deposits also exhibit a random pattern of local, and sometimes considerable, variation in thickness. Thus the distribution of sample points need be only approximately regular and in estimating the mean thickness only simple weighting is necessary. In practice, equal weighting can often be applied to thicknesses at all sample points. If, however, there is a distinctly unequal distribution of points, bias is avoided by dividing the sampled area into broad zones, to each of which a value roughly proportional to its area is assigned. This value is then shared between the data points within the zone as the weighting factor.

APPENDIX C

CLASSIFICATION AND DESCRIPTION OF SAND AND GRAVEL

For the purposes of assessing resources of sand and gravel a classification should take account of economically important characteristics of the deposit, in particular the absolute content of fines and the ratio of sand to gravel.

The terminology commonly used by geologists when describing sedimentary rocks (Wentworth, 1922) is not entirely satisfactory for this purpose. For example, Wentworth proposed that a deposit should be described as a 'gravelly sand' when it contains more sand than gravel and there is at least 10 per cent of gravel, provided that there is less than 10 per cent of material finer than sand (less than $\frac{1}{16}$ mm) and coarser than pebbles (more than 64 mm in diameter). Because deposits containing more than 10 per cent fines are not embraced by this system a modified binary classification based on Willman (1942) has been adopted.

When the fines content exceeds 40 per cent the material is not considered to be potentially workable and falls outside the definition of mineral. Deposits which contain 40 per cent fines or less are classified primarily on the ratio of sand to gravel but qualified in the light of the fines content, as follows: less than 10 per cent fines—no qualification; 10 per cent or more but less than 20 per cent fines—'clayey'; 20 to 40 per cent fines—'very clayey'.

The term 'clay' (as written, with single quote marks) is used to describe all material passing $\frac{1}{16}$ mm. Thus it has no mineralogical significance and includes particles falling within the size range of silt. The normal meaning applies to the term clay where it does not appear in single quotation marks.

The ratio of sand to gravel defines the boundaries between sand, pebbly sand, sandy gravel and gravel (at 19:1, 3:1 and 1:1).

Thus it is possible to classify the mineral into one of twelve descriptive categories (see Figure 8). The procedure is as follows:

1 Classify according to ratio of sand to gravel.
2 Describe fines.

For example, a deposit grading 11 per cent gravel, 70 per cent sand and 19 per cent fines is classified as 'clayey' pebbly sand. This short description is included in the borehole log (see Note 10, Appendix D).

Many differing proposals exist for the classification of the grain size of sediments (Atterberg, 1905; Udden, 1914; Wentworth, 1922; Wentworth, 1935; Allen, 1936; Twenhofel, 1937; Lane and others, 1947). As Archer (1970a, b) has emphasised, there is a pressing need for a simple metric scale acceptable to both scientific and engineering interests, for which the class limit sizes correspond closely with certain marked changes in the natural properties of mineral particles. For example, there is an important change in the degree of cohesion between particles at about the $\frac{1}{16}$-mm size, which approximates to the generally accepted boundary between silt and sand. These and other requirements are met by a system based on Udden's geometric scale and a simplified form of Wentworth's terminology (Table 7), which is used in this Report.

The fairly wide intervals in the scale are consistent with the general level of accuracy of the qualitative assessments of the resource blocks. Three sizes of sand are recognised, fine (+$\frac{1}{16}$–$\frac{1}{4}$ mm), medium (+$\frac{1}{4}$–1 mm) and coarse (+1–4 mm). The boundary at 16 mm distinguishes a range of finer gravel (+4–16 mm), often characterised by abundance of worn tough pebbles of vein quartz, from larger pebbles often of notably different materials. The boundary at 64 mm distinguishes pebbles from cobbles. The term 'gravel' is used loosely to denote both pebble-sized and cobble-sized material.

The size distribution of borehole samples is determined by sieve analysis, which is presented by the laboratory as logarithmic cumulative curves (see, for example, British Standard 1377: 1967). In this report the grading is tabulated on the borehole record sheets (Appendix F), the intercepts corresponding with the simple geometric scale $\frac{1}{16}$ mm, $\frac{1}{4}$ mm, 1 mm, 4 mm, 16 mm and so on as required. Original sample grading curves are available for reference at the appropriate office of the Institute.

Each bulk sample is described, subjectively, by a geologist at the borehole site. Being based on visual examination, the description of the grading is inexact, the accuracy depending on the experience of the observer. The descriptions recorded are modified, as necessary, when the laboratory results become available.

The relative proportions of the rock types present in the gravel fraction are indicated by the use of the words 'and' or 'with'. For example, 'flint and quartz' indicates very approximate equal proportions with neither constituent accounting for less than about 25 per cent of the whole; 'flint with quartz' indicates that flint is dominant and quartz, the principal accessory rock type, comprises 5 to 25 per cent of the whole. Where the accessory material accounts for less than 5 per cent of the whole, but is still readily apparent, the phase 'with some' has been used. Rare constituents are referred to as 'trace'.

The terms used in the field to describe the degree of rounding of particles, which is concerned with the sharpness of the edges and corners of a clastic fragment and not the shape (after Pettijohn, 1957), are as follows.

Angular: showing little or no evidence of wear; sharp edges and corners.

Subangular: showing definite effects of wear. Fragments still have their original form but edges and corners begin to be rounded off.

Subrounded: showing considerable wear. The edges and corners are rounded off to smooth curves. Original grain shape is still distinct.

Rounded: original faces almost completely destroyed, but some comparatively flat surfaces may still remain. All original edges and corners have been smoothed off to rather broad curves. Original shape is still apparent.

Well-rounded: no original faces, edges or corners left. The entire surface consists of broad curves; flat areas are absent. The original shape is suggested by the present form of the grain.

Table 7 Classification of gravel, sand and fines

Size limits	Grain size description	Qualification	Primary classification
64 mm –			
		Coarse	Gravel
16 mm –	Pebble		
		Fine	
4 mm –			
		Coarse	
1 mm –			
	Sand	Medium	Sand
$\frac{1}{4}$ mm –			
		Fine	
$\frac{1}{16}$ mm –			
	Fines (silt and clay)		Fines

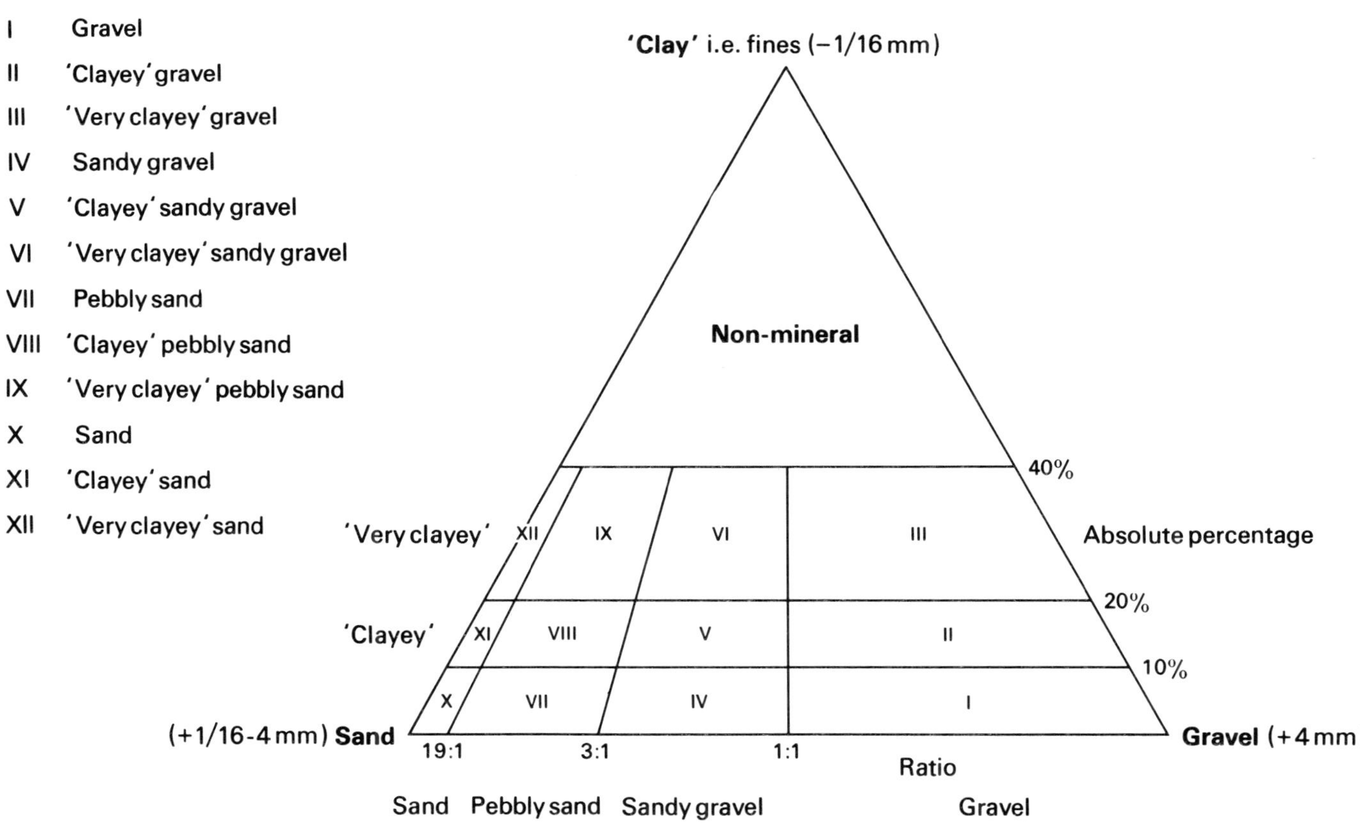

Figure 8 Diagram to show the descriptive categories used in the classification of sand and gravel.

APPENDIX D

EXPLANATION OF THE BOREHOLE RECORDS

TL 60 NE 13[1] **6501 0814**[2] **Roper's Farm, Roxwell**[3] **Block A**

Surface level +44.4 m[4]
Water struck at +39.7 m[5]
Percussion, 152 mm diameter
November 1978[6]

Overburden[7] 1.4 m
Mineral 1.2 m
Waste 4.3 m
Mineral 12.8 m
Bedrock 0.2 m+[8]

LOG

Geological classification[9]	Lithology[10]	Thickness m	Depth m
	Topsoil	0.2	0.2
Boulder Clay	Clay, silty, sandy with pebbles of flint and quartz, pale brown soft	0.3	0.5
	Clay, silty, sandy with pebbles of flint and chalk and a trace of quartz, pale yellow brown, firm	0.9	1.4
Glacial Sand and Gravel	**a** 'Very clayey' gravel Gravel: fine with coarse, angular flint with well rounded chalk and flint, with some subrounded to rounded quartzite, quartz and sandstone, with a trace of limestone, igneous, metamorphic, fossil debris, phosphatic nodules and ironstone Sand: medium and coarse with some fine, predominantly subrounded quartz with coarse and medium angular flint and rounded chalk, yellow brown	1.2	2.6
Boulder Clay	Clay, very sandy, silty with a 'very clayey' sandy chalk gravel between 2.8 and 2.9 m, yellow brown soft	3.4	6.0
	Clay, silty, sandy with pebbles of flint and chalk and many thin sandy silt seams, grey, soft	0.9	6.9
Glacial Sand and Gravel	**b** Gravel: with two thin seams of chalky, flinty, silty brown clay developed between 17.3 and 17.7 m and 19.1 and 19.2 m Gravel: fine with coarse and a trace of cobble grade towards base, angular flint with well rounded chalk and flint with some subrounded to rounded quartzite, quartz and sandstone with a trace of limestone, igneous, metamorphic, fossil debris, phosphatic nodules and ironstone. Chalk content decreases towards base Sand: medium with coarse with some fine predominantly subrounded quartz with coarse and medium angular flint and well rounded chalk, pale brown	12.8	19.7
London Clay	Clay, silty, becomes waxy dark grey brown	0.2+	19.9

GRADING[11]

	Mean for deposit *percentages*			Depth below surface (m)	*percentages*						
	Fines	Sand	Gravel		Fines	Sand			Gravel		
					$-\frac{1}{16}$	$+\frac{1}{16}-\frac{1}{4}$	$+\frac{1}{4}-1$	+1–4	+4–16	+16–64	+64
a	20	37	43	1.4–2.6	20	6	18	13	27	16	0
b	5	44	57	6.9–7.9*	6	6	29	15	26	18	0
				7.9–8.9*	3	6	34	16	27	12	0
				8.9–9.9*	5	4	35	14	29	13	0
				9.9–10.9*	3	4	30	13	32	18	0
				10.9–11.9*	4	4	35	11	26	20	0
				11.9–12.9*	5	3	26	15	35	26	0
				12.9–13.9*	5	3	19	14	31	26	2
				13.9–14.9*	6	3	18	19	32	22	0
				14.9–15.9*	3	2	19	14	27	33	2
				15.9–17.3*	4	1	34	15	29	14	3
				17.7–19.1*	6	3	24	16	32	16	3
				19.2–19.7*	4	3	11	9	31	43	0
				Mean	5	3	27	14	30	20	1
a+b	6	44	50	Mean	6	4	26	14	29	20	1

COMPOSITION[12]

Depth below surface (m)	*Percentage by weight in 4–64 mm fraction* Quartz	Quartz-ite	Sand-stone	Angular flint	Well rounded flint	Chalk	Lime-stone	Igneous and meta-morphic	Fossil debris	Iron-stone	Phos-phatic nodules	Others
1.4–2.6, 6.9–19.7	4	5	4	55	13	16	1	1	1	trace	trace	trace

The numbered paragraphs below correspond with the annotations given on the specimen record.

1 Borehole Registration Number
Each Industrial Minerals Assessment Unit (IMAU) borehole is identified by a Registration Number. This consists of two statements.
 1 The number of the 1:25 000 sheet on which the borehole lies, for example TL 60
 2 The quarter of the 1:25 000 sheet on which the borehole lies and the number of the borehole in a series for that quarter, for example NE 13
Thus the full Registration Number is TL 60 NE 13
Usually this is abbreviated to NE 13 in the text.

2 The National Grid reference
All National Grid references in this publication lie within the 100 km square TL unless otherwise stated. Grid references are given to eight figures, accurate to within 10 m for borehole locations. (In the text, six-figure grid references are used for more approximate locations, for example, for farms.)

3 Location
The position of the borehole is generally referred to the nearest named locality on the 1:25 000 base map and the resource block in which it lies is stated.

4 Surface level
The surface level at the borehole site is given in metres above Ordnance Datum. All measurements were made in metres.

5 Groundwater conditions
If groundwater was present the level at which it was encountered is normally given (in metres above Ordnance Datum).

6 Type of drill and date of drilling
Unless otherwise stated, all boreholes were drilled by a shell and auger rig using 6-inch casing. The month and year of completion of the hole are stated.

7 Overburden, Mineral, Waste and Bedrock
Mineral is sand and gravel which, as part of a deposit, falls within the arbitrary definition of potentially workable material (see p. 1). Bedrock is the 'formation', 'country rock' or 'rock head' below which potentially workable sand and gravel will not be found. Waste is any material other than bedrock or mineral. Where waste occurs between the surface and mineral it is classified as overburden.

8 The plus sign (+) indicates that the base of the deposit was not reached during drilling.

9 Geological classification
The geological classification is given whenever possible.

10 Lithological description
When sand and gravel is recorded a general description based on the grading characteristics (for details see Appendix C) is followed by more detailed particulars. The description of other rocks is based on visual examination, in the field. Where more than one mineral deposit is recognised, each is designated by a letter, e.g. a, b, etc.

11 Grading data
A continuous series of bulk samples is taken throughout the thickness of sand and gravel. A new sample is commenced whenever there is an appreciable lithological change or at every 1 m of depth.

For each bulk sample the percentages of fines ($-\frac{1}{16}$ mm), fine sand ($+\frac{1}{16}-\frac{1}{4}$ mm), medium sand ($+\frac{1}{4}-1$ mm), coarse sand (+1–4 mm), fine gravel (+4–16 mm) and coarse gravel (+16 mm) are stated. The mean grading of groups of samples making up an identified mineral horizon are also given in detail and, to the left, in summary. Where more than one horizon is recognised the mean grading for the whole of the mineral in the borehole is also given. Where necessary in calculating the mean grading, data for individual samples are weighted by the thickness represented.

Fully representative sampling of sand and gravel is difficult to achieve, particularly where groundwater levels are high. Comparison between boreholes and adjacent exposures suggests that in borehole samples the proportion of sand may be higher and the proportion of fines and coarse gravel may be lower.

12 Composition
Details of the composition of the gravel fraction of selected samples or grouped samples may be given. Where appropriate the calculated weighted mean composition of grouped samples may be indicated. An asterisk * means that samples were obtained from below the water table using a baling technique.

APPENDIX E

LIST OF IMAU BOREHOLES USED IN THE ASSESSMENT OF RESOURCES

Borehole number	Grid reference*	Block
NW 8	6031 0704	A
9	6023 0559	A
10	6175 0843	A
11	6181 0731	A
12	6183 0617	A
13	6271 0952	A
14	6369 0884	A
15	6303 0742	A
16	6349 0636	A
17	6393 0519	A
18	6428 0738	A
19	6490 0648	A
NE 12	6534 0988	A
13	6501 0814	A
14	6548 0694	A
15	6598 0601	A
16	6516 0564	A
17	6677 0938	A
18	6621 0862	A
19	6638 0742	A
20	6622 0518	A
21	6758 0952	A
22	6743 0816	A
23	6866 0940	A
24	6835 0759	C
25	6817 0548	C
SW 23	6014 0452	B
24	6018 0295	C
25	6132 0128	C
26	6238 0275	B
27	6222 0149	B
28	6385 0026	B
29	6483 0362	B
E1†	6234 0170	B
SE 83	6558 0298	B
84	6632 0358	B
85	6662 0086	C
86	6810 0410	C
87	6877 0376	C
88	6842 0195	C
89	6981 0415	C
90	6983 0282	B
91	6997 0005	B

* All grid references fall within the 100-km square TL.
† Exposure reference number.

APPENDIX F

INDUSTRIAL MINERALS ASSESSMENT UNIT BOREHOLE RECORDS

TL 60 NW 8 6031 0704 Spains Wood, Willingale **Block A**

Surface level +75.8 m
Water not struck
Percussion 152 mm diameter
November 1978

Waste 13.9 m
Bedrock 0.5 m+

LOG

Geological classification	Lithology	Thickness m	Depth m
	Topsoil	0.3	0.3
Boulder Clay	Clay, with flint pebbles, pale brown	0.3	0.6
	Clay, silty, with pebbles of flint, quartz, quartzite and some chalk, light brown mottled grey becoming darker with depth	4.3	4.9
	Clay, silty, with pebbles of flint, quartz and chalk, dark grey, stiff	8.1	13.0
	Clay, silty, with pebbles of flint and quartz, dark brown	0.9	13.9
London Clay	Clay, silty, fine sandy, micaceous, brown-grey	0.5+	14.4

TL 60 NW 9 6023 0559 Hulkes Farm, Willingale **Block A**

Surface level +79.5 m
Water not struck
Percussion 152 mm diameter
December 1978

Waste 16.8 m
Bedrock 0.9 m+

LOG

Geological classification	Lithology	Thickness m	Depth m
	Topsoil	0.2	0.2
Boulder Clay	Clay, silty, with flint pebbles, pale brown mottled grey	0.4	0.6
	Clay, silty, sandy, with pebbles of flint, quartz and chalk, dark brown mottled dark grey	3.4	4.0
	Clay, silty, with pebbles of chalk, flint and quartz, dark grey with some brown mottling	11.9	15.9
	Clay, silty, increasingly pebbly with depth predominantly flint and quartz, dark brown	0.9	16.8
London Clay	Clay, very silty, micaceous, dark grey brown with abundant pyritised remains	0.9+	17.7

TL 60 NW 10 6175 0843 Bushey Heys Spring, Roxwell **Block A**

Surface level +68.0 m
Water struck at +61.4 m
Percussion 152 mm diameter
November 1978

Overburden 5.4 m
Mineral 6.6 m
Bedrock 0.6 m+

LOG

Geological classification	Lithology	Thickness m	Depth m
	Topsoil	0.1	0.1
Boulder Clay	Clay, silty, light brown soft	0.3	0.3
	Clay, silty, with pebbles of chalk, flint and quartz, light brown mottled grey becoming darker with depth	1.9	2.2
	Clay, silty, with chalk and flint pebbles, dark grey	3.0	5.2
	Clay, very sandy, silty, with flint pebbles, orange brown	0.2	5.4
Kesgrave Sands and Gravels	'Clayey' sandy gravel, 'very clayey' pebbly sand between 8.3 m and 11.6 m with fine clay laminae Gravel: fine and coarse, with a trace of cobble grade at base, predominantly angular and well rounded flint, with rounded quartz and quartzite, with some sandstone, and a trace of igneous and metamorphic material Sand: medium with some coarse and fine grade, predominantly subrounded quartz with coarse angular flint, orange brown	6.6	12.0
London Clay	Clay, silty, faintly laminated, micaceous, rusty brown	0.3	12.3
	Clay, silty, micaceous, dark blue-grey, firm	0.3+	12.6

GRADING

Mean for deposit *percentages*			Depth below surface (m)	*percentages*						
Fines	Sand	Gravel		Fines	Sand			Gravel		
				$-\frac{1}{16}$	$+\frac{1}{16}-\frac{1}{4}$	$+\frac{1}{4}-1$	+1–4	+4–16	+16–64	+64
14	56	30	5.4–6.4	14	7	44	6	21	8	0
			6.4–7.4*	2	3	37	10	34	14	0
			7.4–8.3*	4	2	22	15	37	20	0
			8.3–11.6*	21	7	57	2	7	6	0
			11.6–12.0*	3	3	26	9	31	25	3
			Mean	14	5	45	6	19	11	0

COMPOSITION

Depth below surface (m)	*Percentage by weight in* 4–64 *mm fraction*					
	Quartz	Quartzite	Sandstone	Angular flint	Well rounded flint	Igneous and metamorphics
5.4–12.0	16	12	4	46	21	1

TL 60 NW 11 6181 0731 Sandpit Wood, Willingale **Block A**

Surface level +72.1 m
Water struck at +60.5 m
Percussion 152 mm diameter
November 1978

Overburden 10.4 m
Mineral 3.8 m
Bedrock 0.3 m+

LOG

Geological classification	Lithology	Thickness m	Depth m
	Topsoil	0.1	0.1
	Subsoil, sandy, silty clay, with chalk and flint pebbles, pale brown	0.6	0.7
Boulder Clay	Clay, silty, with a seam of putty chalk 5 cm thick at 0.9 m, pale brown	0.3	1.0
	Clay, silty, chalky, pebbly, pale brown mottled blue grey becoming dark blue grey with depth	3.1	4.1
	Clay, silty, with pebbles of chalk and flint and some cobble grade flint, dark grey	3.6	7.7
	Clay, silty, sandy, with flint and a few chalk pebbles, blue grey	1.8	9.5
	Clay, very fine sandy, pebbly, yellow brown	0.9	10.4
Kesgrave Sands and Gravels	Gravel: Gravel: fine with coarse, angular and well-rounded flint, with rounded quartz and quartzite, with some sandstone and igneous and metamorphic material Sand: medium and coarse and a trace of fine, predominantly angular flint and rounded quartz, pale brown	3.8	14.2
London Clay	Clay, silty, faintly laminated, micaceous, blue-grey with a brown tinge	0.3+	14.5

GRADING

Mean for deposit *percentages*			Depth below surface (m)	*percentages*					
Fines	Sand	Gravel		Fines	Sand			Gravel	
				$-\frac{1}{16}$	$+\frac{1}{16}-\frac{1}{4}$	$+\frac{1}{4}-1$	+1–4	+4–16	+16–64
1	38	61	10.4–11.4	0	1	17	16	40	26
			11.4–12.4*	2	3	33	13	31	18
			12.4–13.4*	0	0	15	22	37	26
			13.4–14.2*	2	0	15	18	38	27
			Mean	1	1	20	17	37	24

COMPOSITION

Depth below surface (m)	*Percentage by weight in* 4–64 *mm fraction*						
	Quartz	Quartzite	Sandstone	Angular flint	Well rounded flint	Igneous and metamorphic	Others
10.4–14.2	16	13	5	36	27	3	trace

TL 60 NW 12 6183 0617 Rolls Farm, Willingale **Block A**

Surface level +71.5 m
Water not struck
Percussion 152 mm diameter
November 1978

Waste 9.5 m
Bedrock 4.5 m+

LOG

Geological classification	Lithology	Thickness m	Depth m
	Topsoil	0.2	0.2
Boulder Clay	Clay, silty, sandy, with flint and chalk pebbles and rare quartz, pale brown mottled grey, firm	1.6	1.8
	Clay, very silty, with flint, chalk and quartz pebbles. Thin sand seams throughout, some flint and limestone cobbles, dark grey with a brown tinge, stiff	6.7	8.5
	Clay, very pebbly and sandy, with silt, many pebbles of rounded flint, quartz, quartzite and sandstone, brown, firm	1.0	9.5
? London Clay	'Clayey' sand: medium and fine with a trace of coarse, very rare rounded flint pebbles, fines predominantly silt with some clay. Some thin seams of clear fine white sand throughout, yellow brown	4.5+	14.0

TL 60 NW 13 6271 0952 Mountneys, Roxwell **Block A**

Surface level +52.1 m
Water struck at +40.6 m
Percussion 152 mm diameter
November 1978

Waste 14.5 m
Bedrock 0.4 m+

LOG

Geological classification	Lithology	Thickness m	Depth m
	Topsoil	0.3	0.3
Boulder Clay	Clay, silty, with chalk and flint pebbles, pale brown mottled orange brown	2.1	2.4
	Clay, silty, with chalk pebbles, dark grey mottled orange brown	0.6	3.0
	Chalk, putty, creamy-white	0.4	3.4
	Clay, silty, with pebbles of chalk flint and quartz, dark grey with rare brown mottling, become sandy towards base	8.4	11.8
	Clay and gravel seams, alternation of pale brown very chalky flint gravel and dark grey chalky silty clay	1.1	12.9
Glacial Sand and Gravel	Gravel: becoming sandy towards base Gravel: fine with some coarse, predominantly angular and well rounded flint, with rounded quartz, and some subrounded quartzite and sandstone, with a trace of limestone, igneous, metamorphic, chalk, pyrite and fossil debris Sand: medium with coarse and some fine, predominantly rounded quartz with angular coarse flint and chalk, brown, with some thin clay partings	1.6	14.5
London Clay	Clay, silty, fine sandy, micaceous, dark grey, firm	0.4+	14.9

GRADING

Mean for deposit *percentages*			Depth below surface (m)	*percentages*					
Fines	Sand	Gravel		Fines	Sand			Gravel	
				$-\frac{1}{16}$	$+\frac{1}{16}-\frac{1}{4}$	$+\frac{1}{4}-1$	+1–4	+4–16	+16–64
1	37	62	12.9–13.9*	1	2	13	11	57	16
			13.9–14.5*	2	6	39	11	25	17
			Mean	1	3	23	11	46	16

COMPOSITION

Depth below surface	*Percentage by weight in 4–64 mm fraction*										
	Quartz	Quartz-ite	Sand-stone	Angular flint	Well rounded flint	Chalk	Lime-stone	Igneous and meta-morphic	Pyrite	Fossil debris	Others
12.9–14.5	13	6	4	40	30	1	2	2	1	1	trace

TL 60 NW 14 **6369 0884** **Cooley Spring, Roxwell** **Block A**

Surface level +48.3 m
Water struck at +39.2 m
Percussion 152 mm diameter
November 1978

Waste 11.9 m
Bedrock 1.8 m+

LOG

Geological classification	Lithology	Thickness m	Depth m
	Topsoil	0.3	0.3
Boulder Clay	Clay, silty, with some fine sand, with pebbles of flint quartz and sandstone, and roollets, yellow brown, soft	0.3	0.6
	Clay, silty, with flint and quartz and some chalk pebbles, yellow brown becoming mottled grey	2.1	2.7
	Clay, silty, with flint, quartz and chalk pebbles up to cobble grade. With thin sand seams from 6.1 m. Stiff becoming soft at base, dark blue grey with a brown tinge	6.4	9.1
Kesgrave Sand and Gravels	Gravel: Gravel: fine and coarse, predominantly angular with some well rounded flint with some rounded to subrounded quartz, quartzite and sandstone, and a trace of igneous and metamorphic material. Sand: medium with some coarse and fine predominantly rounded quartz with some coarse angular flint, yellow brown	2.8	11.9
London Clay	Clay, silty, waxy, with some comminuted shale debris and pyrite nodules, brown	1.8+	13.7

GRADING

Mean for deposit *percentages*			Depth below surface (m)	*percentages*					
Fines	Sand	Gravel		Fines	Sand			Gravel	
				$-\frac{1}{16}$	$+\frac{1}{16}-\frac{1}{4}$	$+\frac{1}{4}-1$	+1–4	+4–16	+16–64
1	42	57	9.1–10.1	1	4	29	6	35	25
			10.1–11.1	1	5	41	10	34	9
			11.1–11.9	2	1	20	7	23	47
			Mean	1	3	31	8	31	26

COMPOSITION

Depth below surface (m)	*Percentage by weight in* 4–64 *mm fraction*						
	Quartz	Quartzite	Sandstone	Angular flint	Well rounded flint	Igneous and metamorphic	Others
9.1–11.9	7	7	2	55	28	1	trace

TL 60 NW 15 **6303 0742** **Tye Hall, Roxwell** **Block A**

Surface level +56.0 m
Water struck at +51.9 m
Percussion 152 mm diameter
November 1978

Waste 9.9 m
Bedrock 1.6 m+

LOG

Geological classification	Lithology	Thickness m	Depth m
	Topsoil	0.2	0.2
Boulder Clay	Clay, silty, with flint and some chalk pebbles, brown	0.6	0.8
	Clay, silty, with flint quartz and chalk pebbles, and some carbonaceous patches, pale yellow brown mottled pale grey becoming darker with depth	2.6	3.4
	Clay, silty, with flint and chalk pebbles, dark grey mottled orange brown	0.6	4.0
	Clay, sandy, silty, with some chalk pebbles, dark grey	0.5	4.5
Glacial Sand and Gravel	**a** Very clayey sandy gravel: Gravel: fine and coarse, angular with well rounded flint, with some subrounded to rounded chalk, sandstone, limestone, quartz and quartzite, and a trace of pyrite, fossil debris, igneous, metamorphic and ironstone Sand: medium with coarse and fine, predominantly rounded quartz, with angular corase flint, and medium and coarse well rounded chalk, pale brown	0.9	5.4
Boulder Clay	Clay, silty, with chalk and flint pebbles, dark grey	2.1	7.5
	Clay, silty, with flint and quartz pebbles, orange brown	0.1	7.6
Kesgrave Sands and Gravels	**b** Sandy gravel: with some thin pebbly clay seams Gravel: fine with coarse, angular flint, with well rounded flint and rounded quartz and quartzite, with some sandstone and igneous and metamorphic material Sand: medium with coarse and a trace of fine, predominantly rounded quartz, with some coarse angular flint	2.3	9.9
London Clay	Clay, silty, micaceous, orange brown	0.6	10.5
	Clay, silty, micaceous, dark grey	1.0+	11.5

GRADING

	Mean for deposit *percentages*			Depth below surface (m)	*percentages*					
	Fines	Sand	Gravel		Fines	Sand			Gravel	
					$-\frac{1}{16}$	$+\frac{1}{16}-\frac{1}{4}$	$+\frac{1}{4}-1$	+1–4	+4–16	+16–64
a	22	52	26	4.5–5.4*	22	10	34	8	13	13
b	2	73	25	7.6–8.6	4	2	58	10	18	8
				8.6–9.6	0	1	60	19	17	3
				9.6–9.9	3	1	50	14	19	13
				Mean	2	1	58	14	18	7

COMPOSITION

Depth below surface (m)	*Percentage by weight in 4–64 mm fraction*											
	Quartz	Quartz-ite	Sand-stone	Angular flint	Well rounded flint	Chalk	Lime-stone	Igneous and meta-morphic	Fossil debris	Iron-stone	Pyrite	Others
4.5–5.4	5	3	7	42	23	9	6	1	1	1	1	1
7.6–9.9	11	8	4	57	17	0	0	2	0	0	0	1

TL 60 NW 16 6349 0636 Blows Farm, Roxwell **Block A**

Surface level +66.7 m
Water struck at +62.6 m
Percussion 152 mm diameter
November 1978

Overburden 6.6 m
Mineral 4.7 m
Bedrock 0.9 m+

LOG

Geological classification	Lithology	Thickness m	Depth m
	Topsoil	0.3	0.3
Boulder Clay	Clay, silty, sandy, with pebbles of flint and a trace of chalk, mid brown, soft	0.3	0.6
	Clay, silty, with pebbles of chalk and flint, with a trace of quartz, brown mottled grey, firm	1.4	2.0
	Clay, silty, with pebbles of flint and chalk, grey-brown, stiff	1.9	3.9
	Clay, sandy, silty, with flints, orange brown	0.1	4.0
	Sand, silty, clayey, with flint and chalk pebbles, pale, yellow brown, soft	0.3	4.3
	Clay, very sandy, silty, with pebbles of flint, quartz and chalk, mid-brown, soft	0.7	5.0
	Clay, silty, with pebbles of flint, quartz and chalk, pale grey brown, soft	1.4	6.4
	Clay, very sandy, silty, with pebbles of flint and quartz, pale orange brown, soft	0.2	6.6
Kesgrave Sands and Gravels	Gravel: 'clayey' between 9.6 and 10.6 Gravel: fine and coarse, angular flint with rounded quartzite, well rounded flint and rounded quartz, with some sandstone and igneous and metamorphics Sand: medium with coarse with some fine, predominantly rounded quartz, with coarse angular flint, pale brown becoming yellow brown	4.7	11.3
London Clay	Clay, silty, mottled and streaked brown and orange brown, soft	0.3	11.6
	Clay, silty, with pockets of dark olive green silty fine sand, dark blue grey	0.6+	12.2

GRADING

Mean for deposit *percentages*			Depth below surface (m)	*percentages*					
Fines	Sand	Gravel		Fines	Sand			Gravel	
				$-\frac{1}{16}$	$+\frac{1}{16}-\frac{1}{4}$	$+\frac{1}{4}-1$	+1–4	+4–16	+16–64
6	42	52	6.6–7.6*	6	6	40	7	19	22
			7.6–8.6*	3	2	28	11	28	28
			8.6–9.6*	1	2	15	11	37	34
			9.6–10.6*	14	2	24	7	26	27
			10.6–11.3*	3	4	49	10	19	15
			Mean	6	3	30	9	26	26

COMPOSITION

Depth below surface (m)	*Percentage by weight in* 4–64 *mm fraction*						
	Quartz	Quartzite	Sandstone	Angular flint	Well rounded flint	Igneous and metamorphic	Others
6.6–11.3	13	17	5	45	16	4	trace

TL 60 NW 17   6393 0519   Hawkin Smiths Farm, Highwood   **Block A**

Surface level +73.,7 m
Water struck at +64.8 m and +64.2 m
Percussion 152 mm diameter
November 1978

Waste 12.0 m
Bedrock 3.6 m+

LOG

Geological classification	Lithology	Thickness m	Depth m
	Topsoil	0.3	0.3
Boulder Clay	Clay, silty, sandy, with chalk pellets, pale brown, soft	0.3	0.6
	Clay, silty, sandy, with pebbles of chalk and flint, pale brown, firm	0.4	1.0
	Clay, silty, with chalk and flint pebbles, brown mottled, pale grey, firm	2.5	3.5
	Clay, silty, with pebbles of chalk flint and quartz, grey, hard	5.4	8.9
	Sandy gravel, predominantly quartz and flint rounded fine and coarse gravel, with rounded fine and medium quartz sand, some angular coarse flint sand, grey brown	0.3	9.2
	Clay, silty, with chalk pellets and pebbles of flint and quartz, becomes very silty and pebbly from 10.2 m, dark grey, firm. Thin sandy gravel seams at 9.5 m	1.7	10.9
	Clay, silty, with chalk pellets and pebbles of flint and quartz, dark brown, firm	0.2	11.1
	Clay, silty and sandy, with a trace of flint and quartz, brown, firm	0.9	12.0
? London Clay	Clay, silty, and fine silty sands, with faint laminations becoming well laminated with depth. Clay partings are waxy and sands contain mica. Mid brown with yellow brown partings	3.6+	15.6

TL 60 NW 18 6428 0738 Greenlane Farm, Roxwell **Block A**

Surface level +55.3 m
Water struck at +53.2
Percussion 152 mm diameter
November 1978

Overburden 2.1 m
Mineral 2.5 m
Bedrock 0.9 m+

LOG

Geological classification	Lithology	Thickness m	Depth m
	Made ground	0.3	0.3
Boulder clay	Clay, very silty, with pebbles of flint and quartz, yellow brown, firm	0.5	0.8
	Clay, silty, with pebbles of flint, quartz and chalk, yellow brown mottled grey, firm	0.8	1.6
	Clay, silty, with pebbles of flint and chalk, dark blue grey, soft	0.5	2.1
Glacial Sand and Gravel	Gravel: Gravel: fine and coarse with a trace of cobble, angular with well rounded flint, with some rounded quartzite, quartz and sandstone, with a trace of chalk, igneous metamorphic and fossil debris. Sand: medium and coarse with some fine, predominantly rounded quartz, with some coarse and medium angular flint, and a trace of chalk, yellow brown	2.5	4.6
London Clay	Clay, silty, waxy, brown, soft	0.1	4.7
	Clay, very silty, fine sandy, dark grey	0.8+	5.5

GRADING

Mean for deposit percentages: Fines	Sand	Gravel	Depth below surface (m)	percentages: Fines	Sand			Gravel		
				$-\frac{1}{16}$	$+\frac{1}{16}-\frac{1}{4}$	$+\frac{1}{4}-1$	+1–4	+4–16	+16–64	+64
4	35	61	2.1–3.1*	5	2	14	11	40	28	0
			3.1–4.1*	3	2	16	22	38	17	2
			4.1–4.6*	3	6	25	10	32	24	0
			Mean	4	3	17	15	37	23	1

COMPOSITION

Depth below surface (m)	*Percentage by weight in* 4–64 *mm fraction*: Quartz	Quartzite	Sandstone	Angular flint	Well rounded flint	Chalk	Igneous and metamorphic	Fossil debris
2.1–4.6	8	9	4	49	27	2	1	trace

TL 60 NW 19 **6490 0648** **Moor Hall, Writtle** **Block A**

Surface level +56.1 m
Water struck at +52.3 m
Percussion 152 mm diameter
December 1978

Overburden 1.1 m
Mineral 4.7 m
Bedrock 1.2 m+

LOG

Geological classification	Lithology	Thickness m	Depth m
	Topsoil	0.2	0.2
Boulder Clay	Clay, silty, with a trace of flint pebbles, pale orange brown	0.3	0.5
	Clay, very sandy, silty, with pebbles of flint, orange brown	0.6	1.1
Kesgrave Sands and Gravels	**a** 'Very clayey' pebbly sand: gravelly in the uppermost 1.0 m Gravel: fine with coarse, predominantly angular and well rounded flint, with some rounded quartz, quartzite and sandstone, and a trace of igneous and metamorphic Sand: medium with fine with some coarse, predominantly subrounded to rounded quartz, orange brown banded yellow and orange	2.7	3.8
	b Gravel: Gravel: fine and coarse, predominantly angular and well rounded flint, with some rounded quartz, quartzite and sandstone, and a trace of igneous and metamorphic Sand: medium with some coarse and fine, predominantly subrounded quartz, with some coarse angular flint, yellow brown	2.0	5.8
London Clay	Clay, silty, orange brown	0.5	6.3
	Clay, silty, micaceous, blue grey, firm	0.7+	7.0

GRADING

	Mean for deposit *percentages*			Depth below surface (m)	*percentages*					
	Fines	Sand	Gravel		Fines	Sand			Gravel	
					$-\frac{1}{16}$	$+\frac{1}{16}-\frac{1}{4}$	$+\frac{1}{4}-1$	+1–4	+4–16	+16–64
a	22	62	16	1.1–2.1	22	6	18	11	30	13
				2.1–3.1	23	33	42	2	0	0
				3.1–3.8	21	22	55	2	0	0
				Mean	22	20	37	5	11	5
b	2	38	60	3.8–4.8*	2	3	28	8	31	28
				4.8–5.8*	2	3	23	11	31	30
				Mean	2	3	26	9	31	29
a+b	13	52	35	Mean	13	13	32	7	20	15

COMPOSITION

Depth below surface (m)	*Percentage by weight in 4–64 mm fraction*					
	Quartz	Quartzite	Sandstone	Angular flint	Well rounded flint	Igneous and metamorphic
1.1–5.8	9	7	3	46	34	1

TL 60 NE 12 6534 0988 Boyton Hall, Roxwell **Block A**

Surface level +53.1 m
Water struck at +35.5 m
Percussion 152 mm diameter
November 1978

Waste 18.6 m+

LOG

Geological classification	Lithology	Thickness m	Depth m
	Topsoil	0.2	0.2
Boulder Clay	Clay, silty, with flint pebbles, pale yellow brown, firm	0.5	0.7
	Clay, silty, with flint and chalk pebbles, pale brown mottled orange brown	3.7	4.4
	Putty chalk, cream, soft	0.6	5.0
	Clay, silty, with pebbles of chalk and flint, pale brown mottled orange brown becomes dark brown mottled grey	3.4	8.4
	Clay, silty, very chalky, with flint pebbles, dark grey mottled orange brown	4.8	13.2
	Clay, silty, with pebbles of chalk and flint, dark grey	4.4	17.6
	Interbedded medium to fine very silty chalky sand with silty clay containing chalk pellets, dark grey	1.0+	18.6

TL 60 NE 13 6501 0814 Ropers Farm, Roxwell **Block A**

Surface level +44.4 m
Water struck at +39.7 m
Percussion 152 mm diameter
November 1978

Overburden 1.4 m
Mineral 1.2 m
Waste 4.3 m
Mineral 12.8 m
Bedrock 0.2 m+

LOG

Geological classification	Lithology	Thickness m	Depth m
	Topsoil	0.2	0.2
Boulder Clay	Clay, silty, sandy, with pebbles of flint and quartz, pale brown, soft	0.3	0.5
	Clay, silty, sandy, with pebbles of flint and chalk and a trace of quartz, pale yellow brown, firm	0.9	1.4
Glacial Sand and Gravel	**a** 'Very clayey' gravel Gravel: fine with coarse, angular flint, with well rounded chalk and flint, with some subrounded to rounded quartzite, quartz and sandstone, with a trace of limestone, igneous, metamorphic, fossil debris, phosphatic nodules and ironstone Sand: medium and coarse with some fine, predominantly subrounded quartz, with coarse and medium angular flint and rounded chalk, yellow brown	1.2	2.6
Boulder Clay	Clay, very sandy, silty, with a 'very clayey' sandy chalk gravel between 2.8 and 2.9 m, yellow brown, soft	3.4	6.0
	Clay, silty, sandy, with pebbles of flint and chalk, and many thin sandy silt seams, grey, soft	0.9	6.9
Glacial Sand and Gravel	**b** Gravel: with two thin seams of chalky, flinty, silty, brown clay developed between 17.3 and 17.7 m and 19.1 and 19.2 m Gravel: fine with coarse and a trace of cobble grade towards base, angular flint, with well rounded chalk and flint, with some subrounded to rounded quartzite, quartz and sandstone, with a trace of limestone, igneous, metamorphic, fossil debris, phosphatic nodules and ironstone. Chalk content decreases towards base Sand: medium with coarse with some fine, predominantly subrounded quartz, with coarse and medium angular flint and well rounded chalk, pale brown	12.8	19.7
London Clay	Clay, silty, becomes waxy, dark grey brown	0.2+	19.9

GRADING

	Mean for deposit *percentages*			Depth below surface (m)	*percentages*						
	Fines	Sand	Gravel		Fines	Sand			Gravel		
					$-\frac{1}{16}$	$+\frac{1}{16}-\frac{1}{4}$	$+\frac{1}{4}-1$	+1–4	+4–16	+16–64	+64
a	20	37	43	1.4–2.6	20	6	18	13	27	16	–
b	5	44	51	6.9–7.9*	6	6	29	15	26	18	0
				7.9–8.9*	3	6	34	16	27	12	0
				8.9–9.9*	5	4	35	14	29	13	0
				9.9–10.9*	3	4	30	13	32	18	0
				10.9–11.9*	4	4	35	11	26	20	0
				11.9–12.9*	5	3	26	15	35	26	0
				12.9–13.9*	5	3	19	14	31	26	2
				13.9–14.9*	6	3	18	19	32	22	0
				14.9–15.9*	3	2	19	14	27	33	2
				15.9–17.3*	4	1	34	15	29	14	3
				17.7–19.1*	6	3	24	16	32	16	3
				19.2–19.7*	4	3	11	9	31	42	0
	5	44	51	Mean	5	3	27	14	30	20	1
a+b	6	44	50	Mean	6	4	26	14	29	20	1

COMPOSITION

Depth below surface (m)	*Percentage by weight in 4–64 mm fraction*											
	Quartz	Quartz-ite	Sand-stone	Angular flint	Well rounded flint	Chalk	Lime-stone	Igneous and meta-morphic	Fossil debris	Iron stone	Phos-phatic nodules	Others
1.4–2.6 6.9–19.7	4	5	4	55	13	16	1	1	1	trace	trace	trace

TL 60 NE 14 **6548 0694** **Newney Hall, Writtle** **Block A**

Surface level +51.4 m
Water struck at +44.1 m
Percussion 152 mm diameter
November 1978

Overburden 5.7 m
Mineral 3.2 m
Bedrock 0.9 m+

LOG

Geological classification	Lithology	Thickness m	Depth m
	Made ground	0.5	0.5
Boulder Clay	Clay, silty, with flint pebbles, brown	0.6	1.1
	Clay, silty, with pebbles of chalk and flint, pale brown mottled orange brown becoming dark brown mottled pale grey	3.1	4.2
	Clay, silty, sandy, increasingly pebbly with depth with flint, and decreasingly chalk	1.5	5.7
Kesgrave Sands and Gravels	Gravel: with some thin silty clay seams in uppermost 1.6 m Gravel: fine and coarse, angular flint, with well rounded flint, with some rounded quartz, quartzite and sandstone, with a trace of igneous and metamorphic Sand: medium and coarse with some fine, predominantly subrounded quartz, and some coarse angular flint, orange brown	3.2	8.9
London Clay	Clay, silty, micaceous, with laminations, orange brown	0.5	9.4
	Clay, silty, micaceous, laminated, blue grey, firm	0.4+	9.8

GRADING

Mean for deposit *percentages*			Depth below surface (m)	*percentages*					
Fines	Sand	Gravel		Fines	Sand			Gravel	
				$-\frac{1}{16}$	$+\frac{1}{16}-\frac{1}{4}$	$+\frac{1}{4}-1$	+1–4	+4–16	+16–64
3	35	62	5.7–6.5	8	5	25	13	32	17
			6.5–7.3	5	4	20	16	37	18
			7.3–8.3*	0	2	15	12	37	34
			8.3–8.9*	0	2	9	17	38	34
			Mean	3	3	18	14	36	26

COMPOSITION

Depth below surface (m)	*Percentage by weight in* 4–64 *mm fraction*					
	Quartz	Quartzite	Sandstone	Angular flint	Well rounded flint	Igneous and metamorphic
5.7–8.9	8	8	5	55	23	1

TL 60 NE 15 **6598 0601** **Bush Farm, Writtle** **Block A**

Surface level +53.6 m
Water not struck
Percussion 152 mm diameter
November 1978

Waste 1.4 m
Bedrock 4.8 m+

LOG

Geological classification	Lithology	Thickness m	Depth m
	Topsoil	0.3	0.3
Head	Clay, silty, sandy, with a trace of fine angular flint pebbles	1.1	1.4
London Clay	Clay, silty, with race nodules, purplish brown mottled pale blue grey streaked orange	3.6	5.0
	Clay, silty, brown becoming dark olive grey, stiff	1.2+	6.2

TL 60 NE 16 **6516 0564** **Little Moor Hall, Writtle** **Block A**

Surface level +65.7 m
Water not struck
Percussion 152 mm diameter
November 1978

Waste 8.1 m
Bedrock 1.8 m+

LOG

Geological classification	Lithology	Thickness m	Depth m
	Topsoil	0.2	0.2
Boulder Clay	Clay, with pebbles of flint, pale yellow brown	0.4	0.6
	Clay, silty, with chalk and some flint pebbles, pale yellow brown mottled orange brown	0.9	1.5
	Clay, very sandy, silty, with some flint pebbles, orange brown	0.1	1.6
	Clay, silty, with pebbles of chalk and flint, brown becoming grey mottled orange brown	1.8	3.4
	Clay, silty, with chalk and some flint pebbles, dark blue grey	3.8	7.2
	Clay, silty, with chalk and some flint pebbles, dark grey brown	0.5	7.7
	Clay, silty, with a trace of chalk pebbles and some flint, dark grey brown	0.4	8.1
London Clay	Clay, silty, micaceous, becoming waxy, dark orange brown becoming grey	1.2	9.3
	Clay, silty, micaceous, dark blue grey	0.6+	9.9

TL 60 NE 17 6677 0938 Chignall St James, Chignall **Block A**

Surface level +41.2 m
Water struck at +31.8 m
Percussion 152 mm diameter
November 1978

Overburden 9.4 m
Mineral 3.7 m
Bedrock 0.8 m+

LOG

Geological classification	Lithology	Thickness m	Depth m
	Topsoil	0.2	0.2
Boulder Clay	Clay, silty, with pebbles of chalk and flint, dark brown mottled yellow brown becomes mottled grey with depth	4.2	4.4
	Clay, silty, with pebbles of chalk and some flint, dark grey	3.0	7.4
	Clay, silty with pebbles of flint and a trace of quartzite, dark grey, soft	2.0	9.4
Kesgrave Sands and Gravels	Gravel: Gravel: fine and coarse with a trace of cobble, angular flint with well rounded flint, and some quartz, quartzite and sandstone, with a trace of igneous and metamorphic Sand: medium with some coarse and fine, predominantly subrounded quartz sand, with some angular coarse flint, orange brown	3.7	13.1
London Clay	Clay, silty, micaceous, orange brown becoming grey and blue grey with depth	0.8+	13.9

GRADING

Mean for deposit *percentages*			Depth below surface (m)	*percentages*						
Fines	Sand	Gravel		Fines	Sand			Gravel		
				$-\frac{1}{16}$	$+\frac{1}{16}-\frac{1}{4}$	$+\frac{1}{4}-1$	+1–4	+4–16	+16–64	+64
3	47	50	9.4–10.4*	3	7	48	7	24	11	0
			10.4–11.4*	1	5	31	8	30	25	0
			11.4–12.4*	6	4	30	10	27	21	2
			12.4–13.1*	1	3	21	14	32	29	0
			Mean	3	5	33	9	28	21	1

COMPOSITION

Depth below surface (m)	*Percentage by weight in* 4–64 *mm fraction*					
	Quartz	Quartzite	Sandstone	Angular flint	Well rounded flint	Igneous and metamorphic
9.4–10.4	8	6	2	58	26	trace

TL 60 NE 18 6621 0862 Blackwall Bridge, Roxwell

Block A

Surface level +43.5 m
Water seepage at +42.5 m
Percussion 152 mm diameter
November 1978

Waste 18.9 m
Bedrock 1.6 m+

LOG

Geological classification	Lithology	Thickness m	Depth m
	Topsoil	0.2	0.2
Boulder Clay	Clay, silty, sandy, with some thin discrete sand seams, with pebbles of chalk and flint, pale yellow brown, soft	3.3	3.5
	Clay, silty, with pebbles of chalk and flint with some quartz, red mudstone and black shale, pale grey brown becoming blue grey, firm becoming stiff	14.0	17.5
	Clay, silty, with pebbles of chalk and rounded quartz and flint with some angular flint, dark brown, firm	1.4	18.9
London Clay	Clay, silty, dark brown, firm becoming waxy and fissured	1.6+	20.5

TL 60 NE 19 6638 0742 Reeds Farm, Writtle

Block A

Surface level +34.5 m
Water struck at +33.1 m
Percussion 152 mm diameter
November 1978

Waste 2.0 m
Bedrock 1.3 m+

LOG

Geological classification	Lithology	Thickness m	Depth m
	Topsoil	0.3	0.3
Head	Clay, silty, with some flint and quartz pebbles, brown	1.0	1.3
Kesgrave Sands and Gravels	Gravel Gravel: fine and coarse, angular with well rounded flint, with some rounded quartzite and quartz, with a trace of sandstone, igneous and metamorphic Sand: medium with some coarse and fine, predominantly subrounded quartz, with some coarse angular flint, brown	0.7	2.0
London Clay	Clay, silty, brown	0.6	2.6
	Clay, silty, olive grey	0.7+	3.3

GRADING

Mean for deposit *percentages*			Depth below surface (m)	*percentages*					
Fines	Sand	Gravel		Fines	Sand			Gravel	
				$-\frac{1}{16}$	$+\frac{1}{16}-\frac{1}{4}$	$+\frac{1}{4}-1$	+1–4	+4–16	+16–64
4	41	55	1.3–2.0*	4	3	30	8	31	24

COMPOSITION

Depth below surface (m)	*Percentage by weight in* 4–64 *mm fraction*					
	Quartz	Quartzite	Sandstone	Angular flint	Well rounded flint	Igneous and metamorphic
1.3–2.0	4	9	1	66	20	trace

TL 60 NE 20 **6622 0518** **Halfway House, Writtle** **Block A**

Surface level +65.6 m
Water not struck
Percussion 152 mm diameter
November 1978

Waste 6.5 m
Bedrock 0.8 m+

LOG

Geological classification	Lithology	Thickness m	Depth m
	Topsoil	0.2	0.2
Boulder Clay	Clay, silty, with some pebbles of flint, pale orange grey mottled orange brown	0.5	0.7
	Clay, silty, with pebbles of chalk and flint, pale yellow grey mottled orange brown, becomes stiff	3.6	4.3
	Clay, silty, with some pebbles of flint, dark grey	2.2	6.5
London Clay	Clay, silty, with fine sandy laminations, micaceous, with pyritised remains, waxy purple grey	0.8+	7.3

TL 60 NE 21 **6758 0952** **Britton Hall, Chignall** **Block A**

Surface level +42.0 m
Water struck at +34.9 m
Percussion 152 mm diameter
November 1978

Overburden 5.4 m
Mineral 5.9 m
Bedrock 0.3 m+

LOG

Geological classification	Lithology	Thickness m	Depth m
	Topsoil	0.2	0.2
Boulder Clay	Clay, silty, with flint pebbles, dark brown	0.2	0.4
	Clay, silty, with chalk and flint pebbles, orange brown mottled dark grey with depth	3.6	4.0
	Clay, silty, with chalk and flint pebbles, waxy, blue grey, stiff	1.4	5.4
Kesgrave Sands and Gravels	Sandy gravel: Gravel: fine with coarse and a trace of cobble from 9.4 m, angular with well rounded flint, with some rounded to subrounded quartz, quartzite and sandstone, with a trace of igneous and metamorphic Sand: medium with coarse and some fine, predominantly subrounded quartz, with some angular coarse flint, orange yellow becoming yellow brown	5.9	11.3
London Clay	Clay, silty, micaceous, dark grey brown, stiff	0.3+	11.6

GRADING

Mean for deposit *percentages*			Depth below surface (m)	*percentages*					
Fines	Sand	Gravel		Fines	Sand			Gravel	
				$-\frac{1}{16}$	$+\frac{1}{16}-\frac{1}{4}$	$+\frac{1}{4}-1$	+1–4	+4–16	+16–64
2	52	46	5.4–6.4	6	6	31	12	30	15
			6.4–7.4*	2	2	68	8	14	6
			7.4–8.4*	0	3	22	14	37	24
			8.4–9.4*	1	4	44	7	21	23
			9.4–10.4*	1	4	33	11	35	16
			10.4–11.3*	0	4	27	12	38	19
			Mean	2	4	38	10	29	17

COMPOSITION

Depth below surface (m)	*Percentage by weight in* 4–64 *mm fraction*					
	Quartz	Quartzite	Sandstone	Angular flint	Well rounded flint	Igneous and metamorphic
5.4–11.3	10	7	3	51	28	1

TL 60 NE 22 6743 0816 New Barn, Writtle **Block A**

Surface level +35.3 m
Water level +34.3 m
Percussion 152 mm diameter
November 1978

Waste 3.7 m
Bedrock 4.3 m+

LOG

Geological classification	Lithology	Thickness m	Depth m
	Topsoil	0.2	0.2
Boulder Clay	Clay, silty, with fine angular flint pebbles, rootlets at top, orange brown becoming paler with depth	1.5	1.7
	Clay, silty, with pebbles of flint and quartz, dark brown	2.0	3.7
London Clay	Clay, silty, micaceous, orange brown mottled vivid blue	1.0	4.7
	Clay, silty, fissured, dark blue grey	3.3+	8.0

TL 60 NE 23 6866 0940 Blue House, Chignall **Block A**

Surface level +44.4 m
Water struck at +38.4 m
Percussion 152 mm diameter
November 1978

Overburden 1.9 m
Mineral 7.1 m
Bedrock 1.0 m+

LOG

Geological classification	Lithology	Thickness m	Depth m
	Topsoil	0.2	0.2
Boulder Clay	Clay, sandy, silty, with flint pebbles, dark brown becoming orange brown	1.3	1.5
	Clay, silty, with pebbles of chalk and flint, pale orange brown	0.2	1.7
	Clay, very sandy with flint pebbles, brown	0.2	1.9
Kesgrave Sands and Gravels	**a** Pebbly sand: with thin silty clay seams between 3.9 and 4.9 m Gravel: fine with coarse, angular flint, with well rounded flint and quartz, with some rounded quartzite and subrounded sandstone, with a trace of igneous and metamorphic Sand: medium with some fine and coarse, predominantly subrounded quartz, with a trace of angular coarse flint, orange brown	4.0	5.9
	b Gravel: Gravel: fine and coarse, angular and well rounded flint, with rounded quartz and quartzite, with some subrounded sandstone, and a trace of igneous and metamorphic Sand: medium with coarse and some fine, predominantly subrounded quartz, with angular coarse flint, yellow brown	3.1	9.0
London Clay	Clay, silty, micaceous, with some carbonaceous speckling, pale orange brown with orange very silty seams	0.8	9.8
	Clay, silty, micaceous, waxy, blue grey	0.2+	10.0

GRADING

	Mean for deposit *percentages*			Depth below surface (m)	*percentages*					
	Fines	Sand	Gravel		Fines	Sand			Gravel	
					$-\frac{1}{16}$	$+\frac{1}{16}-\frac{1}{4}$	$+\frac{1}{4}-1$	+1–4	+4–16	+16–64
a	8	80	12	1.9–2.9	6	7	69	3	10	5
				2.9–3.9	6	10	64	3	11	6
				3.9–4.9	11	14	66	4	3	2
				4.9–5.9	7	7	69	4	10	3
				Mean	8	9	67	4	8	4
b	4	37	59	5.9–6.9*	2	3	24	8	30	33
				6.9–7.9*	9	2	17	12	36	25
				7.9–9.0*	1	1	33	10	32	23
				Mean	4	2	25	10	32	27
a+b	6	61	33	Mean	6	6	49	6	19	14

COMPOSITION

Depth below surface (m)	*Percentage by weight in 4–64 mm fraction*					
	Quartz	Quartzite	Sandstone	Angular flint	Well rounded flint	Igneous and metamorphic
1.9–5.9	9	5	4	65	16	1
5.9–9.0	9	9	2	45	34	1

TL 60 NE 24 6835 0759 Warren House, Writtle **Block C**

Surface level +33.3 m
Water struck at +25.8 m and +23.8 m
Percussion 152 mm diameter
November 1978

Overburden 1.4 m
Mineral 2.6 m
Waste 5.6 m
Mineral 3.5 m
Waste 7.9 m+

LOG

Geological classification	Lithology	Thickness m	Depth m
	Topsoil	0.4	0.4
Head	Clay, silty, flint pebbles abundant with depth	0.4	0.8
	Clay, very sandy, silty with pebbles of flint and quartz, with some carbonaceous material, orange brown	0.6	1.4
Glacial Silts	**a** 'Clayey' gravel with disseminated carbonaceous material Gravel: fine with coarse, angular with well rounded flint, with some rounded to subrounded quartz, quartzite and sandstone, with a trace of igneous and metamorphic Sand: medium and coarse with some fine, predominantly subrounded quartz and angular flint, dark brown	2.6	4.0
	Peat, silty, clayey, with abundant shell debris, dark brown and black	1.0	5.0
	Clay, pale yellow grey, soft	0.1	5.1
	Clay, silty, sandy, with many chalk pellets and gastropod shells, grey	0.4	5.5
	Peat, silty, very shelly, greenish, with a thin seam of shell debris at 6.1 m	1.5	7.0
	Peat, silty, sandy, shelly, blue grey, with a thin seam of shell debris at 7.5 m	0.6	7.6
Boulder Clay	Clay, very silty, with pebbles of flint, chalk and quartz, dark blue grey	2.0	9.6
Glacial Sand and Gravel	**b** 'Clayey' pebbly sand: with a seam of pebbly orange brown clay between 12.5 and 12.6 Gravel: fine and coarse, angular flint, with well rounded chalk, with some subangular sandstone, well rounded flint and subrounded limestone, and a trace of quartz, quartzite, igneous, metamorphic, fossils, phosphatic nodules and ironstone Sand: medium with fine and some coarse, predominantly subrounded quartz, with well rounded chalk, grey becoming orange brown	3.5	13.1
Boulder Clay	Clay, silty, very chalky, becoming dark brown with depth	0.3	13.4
	Clay, silty, with pebbles of chalk and flint and some thin seams of silty sand, dark blue grey	7.6+	21.0

GRADING

	Mean for deposit *percentages* Fines	Sand	Gravel	Depth below surface (m)	*percentages* Fines $-\frac{1}{16}$	Sand $+\frac{1}{16}-\frac{1}{4}$	$+\frac{1}{4}-1$	$+1-4$	Gravel $+4-16$	$+16-64$
a	13	39	48	1.4–2.4	18	5	23	11	27	16
				2.4–3.5	12	5	23	9	29	22
				3.5–4.0	7	4	20	16	35	18
				Mean	13	5	23	11	29	19
b	14	66	20	9.6–10.6*	10	19	61	6	4	0
				10.6–11.6*	20	24	40	6	9	1
				11.6–12.5*	14	26	26	7	13	14
				12.6–13.1*	12	11	18	6	14	39
				Mean	14	21	39	6	10	10
a+b	14	54	32	Mean	14	14	32	8	18	14

COMPOSITION

Depth below surface (m)	*Percentage by weight in 4–64 mm fraction* Quartz	Quartzite	Sandstone	Angular flint	Well rounded flint	Chalk	Limesone	Igneous and metamorphic	Fossil debris	Phosphatic nodules	Ironstone	Others
1.4–4.0	4	4	4	63	25	0	0	1	0	0	0	trace
9.6–13.1	1	1	8	58	4	21	4	1	1	1	trace	1

TL 60 NE 25 6817 0548 Writtle Lodge, Writtle **Block C**

Surface level +38.7 m
Water struck at +33.5 m
Percussion 152 mm diameter
December 1978

Waste 6.0 m
Bedrock 2.6 m+

LOG

Geological classification	Lithology	Thickness m	Depth m
	Topsoil	0.2	0.2
Head	Clay, silty, sandy, with some flint pebbles, pale orange brown, firm	0.5	0.7
Boulder Clay	Clay, silty, sandy, with some flint and chalk pebbles, pale brown firm	1.1	1.8
	Clay, silty, sandy, with pebbles of flint and chalk, pale brown mottled grey, stiff	1.7	3.5
	Clay, silty, fine sandy, with chalk and flint pebbles, dark grey	1.7	5.2
	Silt, sandy, clayey, with some sand sized chalk pellets, dark grey	0.2	5.4
	Clay, silty, sandy, with chalk and flint pebbles, dark grey	0.6	6.0
London Clay	Clay, silty, micaceous, with calcareous race nodules, dark grey brown	2.6+	8.6

TL 60 SW 23 6014 0452 Norton Heath, Norton Mandeville **Block B**

Surface level +94.3 m
Water not struck
Percussion 152 mm diameter
November 1978

Overburden 0.2 m
Mineral 3.1 m
Bedrock 1.1 m+

LOG

Geological classification	Lithology	Thickness m	Depth m
	Topsoil	0.2	0.2
Older Head	'Clayey' gravel: becomes more 'clayey' and less pebbly with depth, gravel closely packed in clay matrix Gravel: coarse with fine, predominantly well rounded black, brown, grey blue and red with some angular brown flint, with a trace of quartz, quartzite, igneous, metamorphic and sandstone Sand: medium with some coarse and fine, predominantly rounded quartz and angular flint, yellow brown mottled and streaked orange brown	3.1	3.3
Claygate Beds	Sand, fine with medium, very silty, with seams of clayey white fine sand, yellow brown	1.1+	4.4

GRADING

Mean for deposit *percentages*			Depth below surface (m)	*percentages*					
Fines	Sand	Gravel		Fines	Sand			Gravel	
				$-\frac{1}{16}$	$+\frac{1}{16}-\frac{1}{4}$	$+\frac{1}{4}-1$	+1–4	+4–16	+16–64
12	35	53	0.2–1.2	9	2	10	5	18	56
			1.2–2.2	10	3	19	8	19	41
			2.2–3.3	16	7	44	4	10	19
			Mean	12	4	25	6	15	38

COMPOSITION

Depth below surface (m)	*Percentage by weight in 4–64 mm fraction*					
	Quartz	Quartzite	Sandstone	Angular flint	Well rounded flint	Igneous and metamorphic
0.2–3.3	trace	trace	trace	19	81	trace

TL 60 SW 24 6018 0295 1 km South of Norton Heath, High Ongar **Block C**

Surface level +81.3 m
Water struck at +79.3 m
Percussion 152 mm diameter
November 1978

Overburden 2.0 m
Mineral 1.2 m
Waste 1.3 m
Bedrock 0.7 m+

LOG

Geological classification	Lithology	Thickness m	Depth m
	Topsoil	0.2	0.2
Boulder Clay	Clay, very sandy, silty, with pebbles of angular flint and rounded quartz, dark yellow brown, soft	0.3	0.5
	Clay, silty, sandy, with pebbles of chalk and flint with some quartz, yellow brown mottled grey becoming dark brown mottled pale blue, firm	1.5	2.0
Glacial Sand and Gravel	Gravel: 'clayey' in uppermost 0.1 m Gravel: coarse and fine with a trace of cobble, angular and well rounded flint, with subrounded quartzite, with some rounded quartz and subangular sandstone, with a trace of igneous, metamorphic, chalk and ironstone Sand: medium with coarse and some fine, predominantly angular flint, with rounded to subrounded quartz, yellow brown	1.2	3.2
Boulder Clay	Clay, silty, with chalk pellets and pebbles and a trace of flint, dark blue grey, firm	0.6	3.8
	Clay, silty, sandy, with rounded and angular flint pebbles and some chalk pellets, dark brown	0.7	4.5
? London Clay	Sand, fine, very clayey and silty, with rare well rounded flint pebbles at top, faintly laminated with fine white sand seams, dark green brown	0.7+	5.2

GRADING

Mean for deposit *percentages*			Depth below surface (m)	*percentages*						
Fines	Sand	Gravel		Fines	Sand			Gravel		
				$-\frac{1}{16}$	$+\frac{1}{16}-\frac{1}{4}$	$+\frac{1}{4}-1$	+1–4	+4–16	+16–64	+64
2	35	63	2.0–3.2	2	3	22	10	29	33	1

COMPOSITION

Depth below surface (m)	*Percentage by weight in* 4–64 *mm fraction*							
	Quartz	Quartzite	Sandstone	Angular flint	Well rounded flint	Chalk	Igneous and metamorphic	Ironstone
2.0–3.2	5	12	2	42	39	trace	trace	trace

TL 60 SW 25 **6132 0128** **Fryerning Wood, Fryerning and Ingatestone** **Block C**

Surface level +84.7 m
Water not struck
Percussion 152 mm diameter
November 1978

Waste 5.6 m
Bedrock 0.4 m+

LOG

Geological classification	Lithology	Thickness m	Depth m
	Topsoil	0.3	0.3
Boulder Clay	Clay, silty, sandy, with angular flint pebbles, pale brown, soft	0.7	1.0
	Clay, silty, fine sandy, with angular flint and rounded chalk pebbles, pale brown mottled grey, firm	2.2	3.2
	Very clayey sandy gravel Gravel: fine and coarse, angular and well rounded flint, with some rounded quartz, and a trace of chalk Sand: fine and medium with some coarse, predominantly subrounded quartz, with some angular flint, yellow brown	0.3	3.5
	Clay, very sandy, silty, with pebbles of flint and chalk, brown, firm	0.5	4.0
	Clay, silty, with pebbles of flint and chalk with a trace of quartz, dark grey, firm	1.6	5.6
? Claygate Beds	Clay, silty, with some comminuted shell debris, dark blue grey, firm	0.4+	6.0

TL 60 SW 26 **6238 0275** **Parsons Spring, Highwood** **Block B**

Surface level +93.9 m
Water struck at +87.3 m
Percussion 152 mm diameter
December 1978

Overburden 1.0 m
Mineral 2.8 m
Waste 2.3 m
Mineral 1.1 m
Bedrock 1.8 m+

LOG

Geological classification	Lithology	Thickness m	Depth m
	Made ground	0.4	0.4
Older Head	Clay, very silty, sandy, pebbly, pale yellow brown	0.6	1.0
	a 'Very clayey' sandy gravel: becomes less pebbly with depth Gravel: coarse and fine, well rounded with some angular flint, with a trace of quartz, quartzite, sandstone, chalk, limestone, igneous, metamorphic and ironstone Sand: medium with some fine and coarse, predominantly subrounded quartz and angular flint, orange brown	2.8	3.8
	Silt, very sandy, with some well rounded flints and a trace of quartz, orange brown streaked white becoming red mottled white and purple	2.3	6.1
	b 'Clayey' pebbly sand, with some thin discrete clay beams Gravel: coarse with fine, well rounded with some angular flint, with a trace of quartz, quartzite sandstone, chalk, limestone, igneous metamorphic and ironstone Sand: medium with some coarse and fine, predominantly angular flint and quartz	1.1	7.2
Claygate Beds	Salt, sandy, orange brown	1.8+	9.0

GRADING

	Mean for deposit *percentages*			Depth below surface (m)	*percentages*					
	Fines	Sand	Gravel		Fines	Sand			Gravel	
					$-\frac{1}{16}$	$+\frac{1}{16}-\frac{1}{4}$	$+\frac{1}{4}-1$	+1–4	+4–16	+16–64
a	21	48	31	1.0–2.0	24	9	27	5	13	23
				2.0–3.0	18	7	26	6	20	23
				3.0–3.8	22	7	57	2	5	7
				Mean	21	8	35	5	13	18
b	16	68	16	6.1–6.9*	16	7	54	7	4	12
				6.9–7.2*		No grading data available				
a+b	20	52	28	Mean	20	8	39	5	11	17

COMPOSITION

Depth below surface (m)	*Percentage by weight in* 4–64 *mm fraction*								
	Quartz	Quartzite	Sandstone	Angular flint	Well rounded flint	Chalk	Limestone	Igneous and metamorphic	Ironstone
1.0–3.8 6.1–7.2	1	1	trace	16	82	trace	trace	trace	trace

TL 60 SW 27 6222 0149 Bell Grove, Fryerning and Ingatestone **Block B**

Surface level +97.6 m
Water not struck
Percussion 152 mm diameter
November 1978

Overburden 0.1 m
Mineral 4.7 m
Bedrock 1.5 m+

LOG

Geological classification	Lithology	Thickness m	Depth m
	Topsoil	0.1	0.1
Older Head	'Clayey' gravel Gravel: fine and coarse, angular with well rounded flint, with some rounded quartzite, quartz and subrounded sandstone, with a trace of igneous and metamorphic Sand: medium with coarse and some fine, predominantly angular flint with subrounded quartz, orange brown	4.7	4.8
Claygate Beds	Clay, very silty, fine sandy, brown mottled pale brown, soft	0.4	5.2
	Interlaminated very clayey silt and silty fine sand, with a waxy red and grey mottled clay, packed with well rounded flints at 5.5 m, yellow brown, brown and white	1.1+	6.3

GRADING

Mean for deposit percentages			Depth below surface (m)	*percentages*					
Fines	Sand	Gravel		Fines	Sand			Gravel	
				$-\frac{1}{16}$	$+\frac{1}{16}-\frac{1}{4}$	$+\frac{1}{4}-1$	+1–4	+4–16	+16–64
10	33	57	0.1–1.1	14	2	18	11	28	27
			1.1–2.1	7	4	32	7	27	23
			2.1–3.1	9	4	15	11	32	29
			3.1–4.1	11	4	14	13	34	24
			4.1–4.8	8	3	24	10	35	20
			Mean	10	3	20	10	31	26

COMPOSITION

Depth below surface (m)	*Percentage by weight in 4–64 mm fraction*					
	Quartz	Quartzite	Sandstone	Angular flint	Well rounded flint	Igneous and metamorphic
0.1–4.8	9	12	3	54	21	1

TL 60 SW 28 **6385 0026** **Fryerning Hall, Ingatestone and Fryerning** **Block B**

Surface level +97.7 m
Water not struck
Percussion 152 mm diameter
December 1978

Overburden 0.1 m
Mineral 3.5 m
Bedrock 0.7 m+

LOG

Geological classification	Lithology	Thickness m	Depth m
	Topsoil	0.1	0.1
Older Head	'Clayey' gravel: becoming less clayey with depth Gravel: coarse with some fine, well rounded with some angular flint, with a trace of quartz, quartzite and sandstone Sand: medium with fine and coarse, predominantly subangular quartz, with some angular flint, brown becoming orange brown mottled green	3.5	3.6
Claygate Beds	Silt, clayey, fine sandy, micaceous, pale orange brown and yellow brown streaked pale green white	0.7+	4.3

GRADING

Mean for deposit *percentages*			Depth below surface (m)	*percentages*					
Fines	Sand	Gravel		Fines	Sand			Gravel	
				$-\frac{1}{16}$	$+\frac{1}{16}-\frac{1}{4}$	$+\frac{1}{4}-1$	+1–4	+4–16	+16–64
14	30	56	0.1–0.6	19	7	12	3	12	47
			0.6–1.6	17	7	16	3	10	47
			1.6–2.6	13	11	21	5	18	32
			2.6–3.6	8	8	17	7	22	39
			Mean	14	8	17	5	16	40

COMPOSITION

Depth below surface (m)	*Percentage by weight in 4–64 mm fraction*					
	Quartz	Quartzite	Sandstone	Angular flint	Well rounded flint	Others
0.1–3.6	1	1	trace	6	92	trace

TL 60 SW 29 **6483 0362** **Writtle Park House, Highwood** **Block B**

Surface level +93.8 m
Water not struck
Percussion 152 mm diameter
November/December 1978

Overburden 0.1 m
Mineral 4.6 m
Bedrock 2.3 m+

LOG

Geological classification	Lithology	Thickness m	Depth m
	Topsoil	0.1	0.1
Older Head	**a** 'Very clayey' gravel Gravel: coarse with fine, well rounded with some angular flint, with a trace of quartz, quartzite, sandstone, igneous and metamorphic Sand: medium with fine and coarse, predominantly subangular quartz, with a trace of angular flint, grey brown mottled orange brown	3.1	3.2
	b 'Very clayey' sand, faintly laminated Gravel: a trace of fine well rounded flint Sand: medium and fine with a trace of coarse, predominantly subangular quartz, yellow brown	1.5	4.7
? Bagshot Beds	Silt, clayey, fine sandy, well laminated, with seams of well rounded flint pebbles, yellow brown and pale grey	2.3+	7.0

GRADING

	Mean for deposit *percentages*			Depth below surface (m)	*percentages*					
	Fines	Sand	Gravel		Fines	Sand			Gravel	
					$-\frac{1}{16}$	$+\frac{1}{16}-\frac{1}{4}$	$+\frac{1}{4}-1$	+1–4	+4–16	+16–64
a	25	29	46	0.1–0.8	21	9	20	4	18	28
				0.8–1.8	44	5	8	3	16	24
				1.8–2.8	15	3	21	8	27	27
				2.8–3.2	12	5	38	3	11	31
				Mean	25	5	19	5	19	27
b	23	77	0	3.2–4.2	19	25	54	2	0	0
				4.2–4.7	30	54	13	2	1	0
				Mean	23	35	40	2	0	0
a+b	24	45	31	Mean	24	15	26	4	13	18

COMPOSITION

Depth below surface (m)	*Percentage by weight in 4–64 mm fraction*						
	Quartz	Quartzite	Sandstone	Angular flint	Well rounded flint	Igneous and metamorphic	Others
0.1–3.2	1	1	trace	15	83	trace	trace
3.2–4.7	Insufficient material						

TL 60 SW E1 **6234 0170** **College Wood, Ingatestone and Fryerning** **Block B**

Surface level C. +95.0 m
Water not struck
Exposure
January 1979

Overburden 0.5 m
Mineral 2.3 m+

LOG

Geological classification	Lithology	Thickness m	Depth m
	Topsoil	0.1	0.1
Older Head	Clay, silty, sandy, with pebbles of well rounded and angular flint, many rootlets, dark brown	0.4	0.5
	'Clayey' gravel Gravel: coarse and fine with some cobble grade, angular with well rounded flint, with some rounded quartz, quartzite and sandstone, and a trace of igneous and metamorphic Sand: medium and coarse with some fine, predominantly angular flint, with rounded quartz, orange brown mottled grey brown, red and yellow brown	2.3+	2.8

GRADING

Mean for deposit *percentages*			Depth below surface (m)	*percentages*						
Fines	Sand	Gravel		Fines	Sand			Gravel		
				$-\frac{1}{16}$	$+\frac{1}{16}-\frac{1}{4}$	$+\frac{1}{4}-1$	+1–4	+4–16	+16–64	+64
16	17	67	0.5–1.5	18	1	5	5	25	43	3
			1.5–2.8	14	4	9	8	26	36	3
			Mean	16	3	7	7	25	39	3

COMPOSITION

Depth below surface (m)	*Percentage by weight in* 4–64 *mm fraction*					
	Quartz	Quartzite	Sandstone	Angular flint	Well rounded flint	Igneous and metamorphic
0.5–2.8	12	8	5	47	27	1

TL 60 SE 83 **6558 0298** **Writtle Park Wood, Highwood** **Block B**

Surface level +88.1 m
Water struck at +81.1 m
Percussion 152 mm diameter
November 1978

Overburden 0.1
Mineral 2.4 m
Bedrock 9.7 m+

LOG

Geological classification	Lithology	Thickness m	Depth m
	Topsoil	0.1	0.1
Older Head	'Very clayey' sandy gravel Gravel: coarse with fine, well rounded with some angular flint, with a trace of quartz, sandstone and quartzite Sand: medium and fine with some coarse, predominantly subangular quartz, with a trace of angular flint, pale brown	2.4	2.5
Claygate Beds	Sand, fine, very silty, clayey, dark orange brown	0.1	2.6
	Interlaminated silty fine sand and sandy silts and clay, clay highly coloured with red, brown and pale grey, yellow brown, white and pale yellow, micaceous	1.6	4.2
	Sand, fine, silty, micaceous, laminated, pale yellow brown	0.2	4.4
	Sand, fine, silty, micaceous, some faint laminations, pale green brown	1.6	6.0
	Clay, silty, waxy, with thin seams of pale pinkish brown silt, dark pinkish brown, with ochre brown pockets	1.0	7.0
	Sand, fine, silty, with some dark green mineral grains, pale green brown	3.1	10.1
	Silt, fine sandy, with more clayey seams, laminated yellow brown, ochre brown and pale yellow brown	1.5	11.6
	Clay, silty, sandy, with oxidised pyrite nodules, orange brown	0.2	11.8
	Clay, silty, with some pyrite nodules, dark blue grey	0.4+	12.2

GRADING

Mean for deposit *percentages*			Depth below surface (m)	*percentages*					
Fines	Sand	Gravel		Fines	Sand			Gravel	
				$-\frac{1}{16}$	$+\frac{1}{16}-\frac{1}{4}$	$+\frac{1}{4}-1$	+1–4	+4–16	+16–64
22	41	37	0.1–0.8	19	15	16	3	20	27
			0.8–1.8	28	20	24	3	8	17
			1.8–2.5	18	17	17	3	11	34
			Mean	22	18	20	3	12	25

COMPOSITION

Depth below surface (m)	*Percentage by weight in 4–64 mm fraction*				
	Quartz	Quartzite	Sandstone	Angular flint	Well rounded flint
0.1–2.5	1	trace	1	10	88

TL 60 SE 84 **6632 0358** **Bakers Wood, Margaretting** **Block B**

Surface level +82.6 m
Water not struck
Percussion 152 mm diameter
December 1978

Waste 2.1 m
Bedrock 3.9 m+

LOG

Geological classification	Lithology	Thickness m	Depth m
	Topsoil	0.3	0.3
Older Head	Clay, very silty, sandy, with some rounded flints, pale grey and orange brown	1.8	2.1
(?) Bagshot Beds	Silt, fine sandy, with thin clay seams, micaceous, orange brown	2.3	4.4
(?) Claygate Beds	Clay, silty, micaceous, blue grey	1.6+	6.0

TL 60 SE 85 **6662 0086** **Canterburys, Margaretting** **Block C**

Surface level +53.3 m
Water struck at +51.3 m
Percussion 152 mm diameter
December 1978

Overburden 0.1 m
Mineral 3.3 m
Bedrock 0.9 m+

LOG

Geological classification	Lithology	Thickness m	Depth m
	Topsoil	0.1	0.1
Head	Gravel: clayey in uppermost 1.6 m Gravel: fine and coarse with a trace of cobble, well rounded red, black and brown and angular flint, with some rounded quartz and quartzite, with a trace of sandstone, igneous and metamorphic Sand: medium with coarse and fine, predominantly subangular quartz, with some angular flint, brown	3.3	3.4
(?) Claygate Beds	Clay, silty, with oxidised pyrite nodules, brown mottled orange brown	0.5	3.9
	Clay, silty, micaceous, with pyrite nodules, drab olive green	0.4+	4.3

GRADING

Mean for deposit *percentages*			Depth below surface (m)	*percentages*						
Fines	Sand	Gravel		Fines	Sand			Gravel		
				$-\frac{1}{16}$	$+\frac{1}{16}-\frac{1}{4}$	$+\frac{1}{4}-1$	+1–4	+4–16	+16–64	+64
9	45	46	0.1–0.6	16	4	12	9	29	26	4
			0.6–1.7	11	4	17	8	30	30	0
			1.7–2.7*	5	7	43	7	24	14	0
			2.7–3.4*	5	7	54	9	16	9	0
			Mean	9	5	32	8	25	20	1

COMPOSITION

Depth below surface (m)	*Percentage by weight in 4–64 mm fraction*					
	Quartz	Quartzite	Sandstone	Angular flint	Well rounded flint	Igneous and metamorphic
0.1–3.4	8	5	1	40	46*	trace

* Including 10% 'jasperised' flint.

TL 60 SE 86 6810 0410 Home Farm, Writtle **Block C**

Surface level +57.6 m
Water not struck
Percussion 152 mm diameter
December 1978

Waste 7.7 m
Bedrock 0.7 m+

LOG

Geological classification	Lithology	Thickness m	Depth m
	Topsoil	0.2	0.2
Boulder Clay	Clay, silty, with fine patinated white flint pebbles, yellow brown, firm	0.4	0.6
	Clay, silty, with pebbles of chalk and flint with a trace of quartz, yellow brown mottled grey becomes dark brown mottled blue grey, stiff	2.7	3.3
	Clay, silty, with some thin sand seams, with pebbles of rounded flint and quartz with some angular flints and calcareous nodules	2.5	5.8
	Clay, silty, with many pebbles of well rounded flint, becomes micaceous and less pebbly from 7.0 m, dark grey brown, firm	1.9	7.7
London Clay	Clay, silty, fine sandy, with pyrite nodules and pockets of olive green fine sand, drab olive green, firm	0.7+	8.4

TL 60 SE 87 6877 0376 Elm Farm, Margaretting **Block C**

Surface level +45.2 m
Water not struck
Percussion 152 mm diameter
December 1978

Waste 1.5 m
Bedrock 4.4 m+

LOG

Geological classification	Lithology	Thickness m	Depth m
	Topsoil	0.2	0.2
Head	Clay, sandy, very pebbly, mid brown	0.7	0.9
	Clay, silty, sandy, with some flint pebbles, pale brown mottled orange brown	0.6	1.5
London Clay	Clay, silty, fine sandy, brown mottled orange brown	2.0	3.5
	Septarian nodule, calcareous mudstone, pale grey	0.1	3.6
	Clay, silty, fine sandy, with mica, pyrite nodules and worm burrowed claystone, brown mottled grey becomes dark grey	2.3+	5.9

TL 60 SE 88 **6842 0195** **Whitesbridge Farm, Margaretting** **Block C**

Surface level +38.1 m
Water struck at +35.4 m
Percussion 152 mm diameter
December 1978

Overburden 2.7 m
Mineral 1.2 m
Bedrock 0.8 m+

LOG

Geological classification	Lithology	Thickness m	Depth m
	Topsoil	0.1	0.1
Head	Clay, silty, sandy, with some angular flint pebbles, pale yellow brown, firm	1.3	1.4
	Clay, silty, very sandy, with black carbonaceous pockets, yellow brown mottled pale grey	1.3	2.7
	Sandy gravel: Gravel: fine with coarse and some cobble, angular with well rounded flint, with some rounded quartz and quartzite, and a trace of igneous, metamorphic and sandstone. Sand: medium with coarse with some fine, predominantly rounded quartz, and coarse and medium angular flint, yellow brown	1.2	3.9
London Clay	Clay, silty, brown mottled orange brown	0.5	4.4
	Clay, silty, fine sandy, with pockets of olive green fine sand, drab olive grey	0.3+	4.7

GRADING

Mean for deposit percentages			Depth below surface (m)	percentages						
Fines	Sand	Gravel		Fines	Sand			Gravel		
				$-\frac{1}{16}$	$+\frac{1}{16}-\frac{1}{4}$	$+\frac{1}{4}-1$	+1–4	+4–16	+16–64	+64
8	54	38	2.7–3.9*	8	6	31	17	28	8	2

COMPOSITION

Depth below surface (m)	*Percentage by weight in* 4–64 *mm fraction*					
	Quartz	Quartzite	Sandstone	Angular flint	Well rounded flint	Igneous and metamorphic
2.7–3.9	5	4	trace	61	29	1

TL 60 SE 89 6981 0415 Moulsham Thrift Wood, Chelmsford **Block C**

Surface level +55.2 m
Water struck at +47.5 m
Percussion 152 mm diameter
December 1978

Waste 7.9 m
Bedrock 0.6 m+

LOG

Geological classification	Lithology	Thickness m	Depth m
	Topsoil	0.2	0.2
Head	Clay, very silty, sandy with pebbles of angular and rounded flint and some quartz, brown, firm	0.6	0.8
	Clay, silty, with pebbles of white patinated angular flint and rounded quartz, pale orange brown mottled grey brown and red brown along fissures, very hard	6.2	7.0
	Clay, silty, sandy, with pebbles of flint, quartz and quartzite, dark grey brown, firm	0.7	7.7
	Clay, very sandy, pebbly, brown	0.2	7.9
London Clay	Clay, silty, fine sandy, mucaceous, brown mottled yellow brown, soft	0.3	8.2
	Clay, silty, fine sandy, micaceous, with pockets of olive green fine sand, drab olive grey	0.3+	8.5

TL 60 SE 90 6983 0282 Goat Hall, Chelmsford **Block B**

Surface level +70.8 m
Water not struck
Percussion 152 mm diameter
December 1978

Overburden 0.3 m
Mineral 1.2 m
Bedrock 1.8 m+

LOG

Geological classification	Lithology	Thickness m	Depth m
	Topsoil	0.3	0.3
Older Head	'Very clayey' sand Gravel: coarse and fine, well rounded with angular flint, with some rounded quartz and sandstone Sand: fine with a trace of medium and coarse, predominantly subangular quartz, pale yellow grey	1.2	1.5
Claygate Beds	Sand, very fine, silty, clayey, micaceous, yellow brown mottled grey green and white	1.8+	3.3

GRADING

Mean for deposit *percentages*			Depth below surface (m)	*percentages*					
Fines	Sand	Gravel		Fines	Sand			Gravel	
				$-\frac{1}{16}$	$+\frac{1}{16}-\frac{1}{4}$	$+\frac{1}{4}-1$	+1–4	+4–16	+16–64
29	68	3	0.3–1.5	29	63	3	2	1	2

COMPOSITION

Depth below surface (m)	*Percentage by weight in* 4–64 *mm fraction*			
	Quartz	Sandstone	Angular flint	Well rounded flint
0.3–1.5	3	2	26	69

TL 60 SE 91 6997 0005 Forest Wood, Stock **Block B**

Surface level +88.3 m
Water not struck
Percussion 152 mm diameter
December 1978

Waste 1.4 m
Bedrock 1.6 m+

LOG

Geological classification	Lithology	Thickness m	Depth m
	Topsoil	0.2	0.2
Older Head	Very clayey gravel: Gravel: fine and coarse, well rounded flint, with some angular flint and rounded quartz, with a trace of ironstone, quartzite, sandstone, igneous and metamorphic. Sand: medium with some fine and coarse, predominantly subrounded quartz, pale brown	0.6	0.8
	Clay, silty, yellow brown, mottled orange brown and grey green	0.2	1.0
	'Clayey' pebbly sand: pebbly at base only. Gravel: cobbles with coarse and fine, well rounded flint, with some angular flint and rounded quartz. Sand: medium with fine and some coarse, finely laminated, subangular quartz, orange brown and white	0.4	1.4
? Bagshot Beds	Clay, very silty and sandy, yellow brown mottled pale grey, with fine brown angular iron nodules, lilac tinge in parts	1.6+	3.0

GRADING

Mean for deposit percentages			Depth below surface (m)	*percentages*						
Fines	Sand	Gravel		Fines	Sand			Gravel		
				$-\frac{1}{16}$	$+\frac{1}{16}-\frac{1}{4}$	$+\frac{1}{4}-1$	+1–4	+4–16	+16–64	+64
25	43	32	0.2–0.8	29	5	21	5	22	18	–
			1.0–1.4	18	17	42	3	2	4	14
			Mean	25	10	29	4	14	12	6

COMPOSITION

Depth below surface (m)	*Percentage by weight in* 4–64 *mm fraction*						
	Quartz	Quartzite	Sandstone	Angular flint	Well rounded flint	Igneous and metamorphic	Ironstone
0.2–0.8 } 1.0–1.4 }	2	trace	trace	9	88	trace	1

APPENDIX G

CONVERSION TABLE, METRES TO FEET (to nearest 0.5 ft)

m	ft	m	ft	m	ft	m	ft	m	ft
0.1	0.5	6.1	20	12.1	39.5	18.1	59.5	24.1	79
0.2	0.5	6.2	20.5	12.2	40	18.2	59.5	24.2	79.5
0.3	1	6.3	20.5	12.3	40.5	18.3	60	24.3	79.5
0.4	1.5	6.4	21	12.4	40.5	18.4	60.5	24.4	80
0.5	1.5	6.5	21.5	12.5	41	18.5	60.5	24.5	80.5
0.6	2	6.6	21.5	12.6	41.5	18.6	61	24.6	80.5
0.7	2.5	6.7	22	12.7	41.5	18.7	61.5	24.7	81
0.8	2.5	6.8	22.5	12.8	42	18.8	61.5	24.8	81.5
0.9	3	6.9	22.5	12.9	42.5	18.9	62	24.9	81.5
1.0	3.5	7.0	23	13.0	42.5	19.0	62.5	25.0	82
1.1	3.5	7.1	23.5	13.1	43	19.1	62.5	25.1	82.5
1.2	4	7.2	23.5	13.2	43.5	19.2	63	25.2	82.5
1.3	4.5	7.3	24	13.3	43.5	19.3	63.5	25.3	83
1.4	4.5	7.4	24.5	13.4	44	19.4	63.5	25.4	83.5
1.5	5	7.5	24.5	13.5	44.5	19.5	64	25.5	83.5
1.6	5	7.6	25	13.6	44.5	19.6	64.5	25.6	84
1.7	5.5	7.7	25.5	13.7	45	19.7	64.5	25.7	84.5
1.8	6	7.8	25.5	13.8	45.5	19.8	65	25.8	84.5
1.9	6	7.9	26	13.9	45.5	19.9	65.5	25.9	85
2.0	6.5	8.0	26	14.0	46	20.0	65.5	26.0	85.5
2.1	7	8.1	26.5	14.1	46.5	20.1	66	26.1	85.5
2.2	7	8.2	27	14.2	46.5	20.2	66.5	26.2	86
2.3	7.5	8.3	27	14.3	47	20.3	66.5	26.3	86.5
2.4	8	8.4	27.5	14.4	47	20.4	67	26.4	86.5
2.5	8	8.5	28	14.5	47	20.5	67.5	26.5	87
2.6	8.5	8.6	28	14.6	48	20.6	67.5	26.6	87.5
2.7	9	8.7	28.5	14.7	48	20.7	68	26.7	87.5
2.8	9	8.8	29	14.8	48.5	20.8	68	26.8	88
2.9	9.5	8.9	29	14.9	49	20.9	68.5	26.9	88.5
3.0	10	9.0	29.5	15.0	49	21.0	69	27.0	88.5
3.1	10	9.1	30	15.1	49.5	21.1	69	27.1	89
3.2	10.5	9.2	30	15.2	50	21.2	69.5	27.2	89
3.3	11	9.3	30.5	15.3	50	21.3	70	27.3	89.5
3.4	11	9.4	31	15.4	50.5	21.4	70	27.4	90
3.5	11.5	9.5	31	15.5	51	21.5	70.5	27.5	90
3.6	12	9.6	31.5	15.6	51	21.6	71	27.6	90.5
3.7	12	9.7	32	15.7	51.5	21.7	71	27.7	91
3.8	12.5	9.8	32	15.8	52	21.8	71.5	27.8	91
3.9	13	9.9	32.5	15.9	52	21.9	72	27.9	91.5
4.0	13	10.0	33	16.0	52.5	22.0	72	28.0	92
4.1	13.5	10.1	33	16.1	53	22.1	72.5	28.1	92
4.2	14	10.2	33.5	16.2	53	22.2	73	28.2	92.5
4.3	14	10.3	34	16.3	53.5	22.3	73	28.3	93
4.4	14.5	10.4	34	16.4	54	22.4	73.5	28.4	93
4.5	15	10.5	34.5	16.5	54	22.5	74	28.5	93.5
4.6	15	10.6	35	16.6	54.5	22.6	74	28.6	94
4.7	15.5	10.7	35	16.7	55	22.7	74.5	28.7	94
4.8	15.5	10.8	35.5	16.8	55	22.8	75	28.8	94.5
4.9	16	10.9	36	16.9	55.5	22.9	75	28.9	95
5.0	16.5	11.0	36	17.0	56	23.0	75.5	29.0	95
5.1	17	11.1	36.5	17.1	56	23.1	76	29.1	95.5
5.2	17	11.2	36.5	17.2	56.5	23.2	76	29.2	96
5.3	17.5	11.3	37	17.3	57	23.3	76.5	29.3	96
5.4	17.5	11.4	37.5	17.4	57	23.4	77	29.4	96.5
5.5	18	11.5	37.5	17.5	57.5	23.5	77	29.5	97
5.6	18.5	11.6	38	17.6	57.5	23.6	77.5	29.6	97
5.7	18.5	11.7	38.5	17.7	58	23.7	78	29.7	97.5
5.8	19	11.8	38.5	17.8	58.5	23.8	78	29.8	98
5.9	19.5	11.9	39	17.9	58.5	23.9	78.5	29.9	98
6.0	19.5	12.0	39.5	18.0	59	24.0	78.5	30.0	98.5

REFERENCES

ALLEN, V. T. 1936. Terminology of medium grained sediments. *Rep. Natl. Res. Coun. Washington* 1935–36, *App. 1, Rep. Comm. Sedimentation*, pp. 18–47.

ARCHER, A. A. 1969. Background and problems of an assessment of sand and gravel resources in the United Kingdom. *Proc. 9th Commonw. Min. Metall. Congr., 1969*, Vol. 2: Mining and petroleum geology, pp. 495–508.

— 1970a. Standardisation of the size classification of naturally occurring particles. *Géotechnique*, Vol. 20, pp. 103–107.

— 1970b. Making the most of metrication. *Quarry Managers' J.*, Vol. 54, No. 6, pp. 223–227.

ATTERBERG, A. 1905. Die rationelle Klassifikation der Sande und Kiese. *Chem. Z.*, Vol. 29, pp. 195–198.

BRITISH STANDARD 1377. 1967. *Methods of testing soils for civil engineering purposes.* (London: British Standards Institution.)

BUREAU OF MINES AND GEOLOGICAL SURVEY. 1948. *Mineral Resources of the United States.* (Washington DC: Public Affairs Press.) pp. 14–17.

CRATCHLEY, C. R., and others. 1975. South Essex geological and geotechnical survey. Unpublished report of the Institute of Geological Sciences, for the Department of the Environment Part 2, Geology.

DINES, H. G. and EDMUNDS, F. H. 1925. The geology of the country around Romford. *Mem. Geol. Surv. G.B.*, Sheet 257.

HARRIS, P. M., THURRELL, R. G., HEALING, R. A. and ARCHER, A. A. 1974. Aggregates in Britain. *Proc. R. Soc.*, Ser. A, Vol. 339, pp. 329–353.

LANE, E. W. and others. 1947. Report of the sub-committee on sediment terminology. *Trans. Am. Geophys. Union*, Vol. 28, pp. 936–938.

PETTIJOHN, F. J. 1957. *Sedimentary rocks*, 2nd edition. (London: Harper and Row.)

PRESTWICH, J. 1890. On the relation of the Westleton Shingle to other preglacial drifts in the Thames basin, and on the Southern Drift, with observations on the final elevation and initial subaerial denudation of the Weald: and on the genesis of the Thames. Part III. The Southern Drift. *Q. J. Geol. Soc. London*, Vol. 46, pp. 155–181.

ROSE, J. and ALLEN, P. 1977. Middle Pleistocene stratigraphy in south-east Suffolk. *Q.J. Geol. Soc. London*, Vol. 133, pp. 83–102.

THURRELL, R. G. 1971. The assessment of mineral resources with particular reference to sand and gravel. *Quarry Managers, J.*, Vol. 55, pp. 19–25.

TWENHOFEL, W. H. 1937. Terminology of the fine-grained mechanical sediments. *Rep. Natl Res. Coun. Washington*, 1936–37. *App. 1, Rep. Comm. Sedimentation*, pp. 81–104.

UDDEN, J. A. 1914. Mechanical composition of clastic sediments. *Bull. Geol. Soc. Am.*, Vol. 25, pp. 655–744.

WENTWORTH, C. K. 1922. A scale of grade and class terms for clastic sediments. *J. Geol.*, Vol. 30, pp. 377–392.

— 1935. The terminology of coarse sediments. *Bull. Natl. Res. Coun. Washington*, No. 98, pp. 225–246.

WILLMAN, H. B. 1942. Geology and mineral resources of the Marseilles, Ottawa and Streator quadrangle. *Bull. Illinois State Geol. Surv.*, No. 66, pp. 343–344.

The following reports of the Institute relate particularly to bulk mineral resources

Reports of the Institute of Geological Sciences

Assessment of British Sand and Gravel Resources

1 The sand and gravel resources of the country south-east of Norwich, Norfolk: Resource sheet TG 20. E. F. P. Nickless.
Report 71/20 ISBN 0 11 880216 £1.15

2 The sand and gravel resources of the country around Witham, Essex: Resource sheet TL 81. H. J. E. Haggard.
Report 72/6 ISBN 0 11 880588 6 £1.20

3 The sand and gravel resources of the area south and west of Woodbridge, Suffolk: Resource sheet TM 24. R. Allender and S. E. Hollyer.
Report 72.9 ISBN 0 11 880596 7 £1.70

4 The sand and gravel resources of the country around Maldon, Essex; Resource sheet TL 80. J. D. Ambrose.
Report 73/1 ISBN 0 11 880600 9 £1.20

5 The sand and gravel resources of the country around Hethersett, Norfolk; Resource sheet TG 10. E. F. P. Nickless.
Report 73/4 ISBN 0 11 8806068 £1.60

6 The sand and gravel resources of the country around Terling, Essex: Resource sheet TL 71. C. H. Eaton.
Report 73/5 ISBN 0 11 880608 4 £1.20

7 The sand and gravel resources of the country around Layer Breton and Tolleshunt D'Arcy, Essex: Resource sheet TL 91 and part of TL 90. J. D. Ambrose.
Report 73/8 ISBN 0 11 880614 9 £1.30

8 The sand and gravel resources of the country around Shotley and Felixstowe, Suflolk: Resource sheet TM 23. R. Allender and S. E. Hollyer.
Report 73/13 ISBN 0 11 880625 4 £1.60

9 The sand and gravel resources of the country around Attlebridge, Norfolk: Resource sheet TG 11. E. F. P. Nickless.
Report 73/15 ISBN 0 11 880658 0 £1.85

10 The sand and gravel resources of the country west of Colchester, Essex: Resource sheet TL 92. J. D. Ambrose.
Report 74/6 ISBN 0 11 880671 8 £1.45

11 The sand and gravel resources of the country around Tattingstone, Suffolk: Resource sheet TM 13. S. E. Hollyer.
Report 74/9 ISBN 0 11 880675 0 £1.95

12 The sand and gravel resources of the country around Gerrards Cross, Buckinghamshire: Resource sheet SU 99, TQ 08 and TQ 09. H. C. Squirrell.
Report 74/14 ISBN 0 11 880710 2 £2.20

Mineral Assessment Reports

13 The sand and gravel resources of the country east of Chelmsford, Essex: Resource sheet TL 70. M. R. Clarke.
ISBN 0 11 880744 7 £3.50

14 The sand and gravel resources of the country east of Colchester, Essex: Resource sheet TM 02. J. D. Ambrose.
ISBN 0 11 880745 5 £3.25

15 The sand and gravel resources of the country around Newton on Trent, Lincolnshire: Resource sheet SK 87. D. Price.
ISBN 0 11 880746 3 £3.00

16 The sand and gravel resources of the country around Braintree, Essex: Resource sheet TL 72. M. R. Clarke.
ISBN 0 11 880747 1 £3.50

17 The sand and gravel resources of the country around Besthorpe, Nottinghamshire: Resource sheet SK 86 and part of SK 76. J. R. Gozzard.
ISBN 0 11 880748 X £3.00

18 The sand and gravel resources of the Thames Valley, the country around Cricklade, Wiltshire: Resource sheet SU 09/19 and parts of SP 00/10. P. R. Robson.
ISBN 0 11 880749 8 £3.00

19 The sand and gravel resources of the country south of Gainsborough, Lincolnshire: Resources sheet SK 88 and part of SK 78. J. H. Lovell.
ISBN 0 11 880750 1 £2.50

20 The sand and gravel resources of the country east of Newark upon Trent, Nottinghamshire: Resource sheet SK 85. J. R. Gozzard
ISBN 0 11 880751 X £2.75

21 The sand and gravel resources of the Thames and Kennet Valleys, the country around Pangbourne, Berkshire: Resource sheet SU 67. H. C. Squirrell.
ISBN 0 11 880752 8 £3.25

22 The sand and gravel resources of the country north-west of Scunthorpe, Humberside: Resource sheet SE 81. J. W. C. James.
ISBN 0 11 880753 6 £3.00

23 The sand and gravel resources of the Thames Valley, the country between Lechlade and Standlake: Resource sheet SP 30 and parts of SP 20, SU 29 and SU 39. P. Robson.
ISBN 0 11 881252 1 £7.25

24 The sand and gravel resources of the country around Aldermaston, Berkshire: Resource sheet SU 56 and SU 66. H. C. Squirrell.
ISBN 0 11 881253 X £5.00

25 The celestite resources of the area north-east of Bristol: Resource sheet ST 68 and parts of ST 59, 69, 79, 58, 78, 68 and 77. E. F. P. Nickless, S. J. Booth and P. N. Mosley.
ISBN 0 11 881262 9 £5.00

26 The limestone and dolomite resources of the country around Monyash, Derbyshire: Resource sheet SK 16. F. C. Cox and D. McC. Bridge.
ISBN 0 11 881263 7 £7.00

27 The sand and gravel resources of the country west and south of Lincoln, Lincolnshire: Resource sheets SK 95, SK 96 and SK 97. I. Jackson.
ISBN 0 11 884003 7 £6.00

28. The sand and gravel resources of the country around Eynsham, Oxfordshire: Resource sheet SP 40 and part of SP 41. W. J. R. Harries.
ISBN 0 11 884012 6 £3.00

29 The sand and gravel resources of the country south-west of Scunthorpe, Humberside: Resource sheet SE 80. J. H. Lovell.
ISBN 0 11 884013 4 £3.50

30 Procedure for the assessment of limestone resources. F. C. Cox, D. McC. Bridge and J. H. Hull.
ISBN 0 11 884030 4 £1.25

31 The sand and gravel resources of the country west of Newark upon Trent, Nottinghamshire: Resource sheet SK 75. D. Price and P. J. Rogers.
ISBN 0 11 884031 2 £3.50

32 The sand and gravel resources of the country around Sonning and Henley: Resource sheet SU 77 and SU 78. H. C. Squirrell.
ISBN 0 11 884032 0 £5.25

33 The sand and gravel resources of the country north of Gainsborough: Resource sheet SK 89. J. R. Gozzard and D. Price
ISBN 0 11 884033 9 £4.50

34 The sand and gravel resources of the Dengie Peninsula, Essex: Resource sheet TL 90, etc. M. B. Simmons.
ISBN 0 11 884081 9 £5.00

35 The sand and gravel resources of the country around Darvel: Resource sheet NS 53, 63, etc. E. F. P. Nickless, A. M. Aitken and A. A. McMillan.
ISBN 0 11 884082 7 £7.00

36 The sand and gravel resources of the country around Southend-on-Sea, Essex: Resource sheets TQ 78/79 etc. S. E. Hollyer and M. B. Simmons.
ISBN 0 11 884083 5 £7.50

37 The sand and gravel resources of the country around Bawtry, South Yorkshire: Resource sheet SK 69. A. R. Clayton
ISBN 0 11 884053 3 £5.75

38 The sand and gravel resources of the country around Abingdon, Oxfordshire: Resource sheet SU 49, 59, SP 40, 50. C. E. Corser.
ISBN 0 11 884084 5 £5.50

39 The sand and gravel resources of the Blackwater Valley (Aldershot) area: Resource sheet SU 85, 86, parts SU 84, 94, 95, 96. M. R. Clarke, A. J. Dixon and M. Kubala.
ISBN 0 11 884085 1 £7.00

40 The sand and gravel resources of the country west of Darlington, County Durham: Resource sheet NZ 11, 21. A. Smith.
ISBN 0 11 884086 X £5.00

41 The sand and gravel resources of the country around Garmouth, Grampian Region: Resource sheet NJ 36. A. M. Aitken, J. W. Merritt and A. J. Shaw.
ISBN 0 11 884090 8 £8.75

42 The sand and gravel resources of the country around Maidenhead and Marlow: Resource sheet SU 88, parts SU 87, 97, 98. P. N. Dunkley.
ISBN 0 11 884091 6 £5.00

43 The sand and gravel resources of the country around Misterton, Nottinghamshire: Resource sheet SK 79. D. Thomas and D. Price.
ISBN 0 11 884092 4 £5.25

44 The sand and gravel resources of the country around Sedgefield, Durham: Resource sheet NZ 32. M. D. A. Samuel.
ISBN 0 11 884093 2 £5.75

45 The sand and gravel resources of the country around Brampton, Cumbria: Resource sheet NY 55, part 56. I. Jackson.
ISBN 0 11 884094 0 £6.75

46 The sand and gravel resources of the country around Harlow, Essex: Resource sheet TL 41. P. M. Hopson.
ISBN 0 11 884107 6 £9.50

47 The limestone and dolomite resources of the country around Wirksworth, Derbyshire: Resource sheet SK 25, part 35. F. C. Cox and D. J. Harrison.
ISBN 0 11 884108 4 £15.00

48 The sand and gravel resources of the Loddon Valley area: Sheets SU 75, 76, parts 64, 65, 66 and 74. M. R. Clarke, E. J. Raynor and R. S. Sobey.
ISBN 0 11 884109 2 £8.75

49 The sand and gravel resources of the country around Lanark, Strathclyde Region: Resource sheet NS 94, part 84. J. L. Laxton and E. F. P. Nickless.
ISBN 0 11 884112 2 £11.00

50 The sand and gravel resources of the country around Fordingbridge, Hampshire: Resource sheet SU 11 and parts of SU 00, 01, 10, 20 and 21. M. Kubala.
ISBN 0 11 884111 4 £7.75

51 The sand and gravel resources of the country north of Bournemouth, Dorset: Resource sheet SU 00, 10, 20, SZ 09, 19 and 29. M. R. Clarke.
ISBN 0 11 8841106 *not yet priced*

52 The sand and gravel resources of the country between Hatfield Heath and Great Waltham, Essex: Resource sheet TL 51 and 61. R. J. Marks.
ISBN 0 11 8841130 £8.50

53 The sand and gravel resources of the country around Cottenham, Cambridgeshire: Resource sheets TL 46 and 47. A. J. Dixon.
ISBN 0 11 8841149 £9.25

54 The sand and gravel resources of the country around Huntingdon and St Ives, Cambridgeshire: Resource sheets TL 16, 17, 26, 27, 36 and 37. R. W. Gatliff.
ISBN 0 11 884115 7 *not yet priced*

55 The sand and gravel resources of the country around Ipswich, Suffolk: Resource sheet TM 14. R. Allender and S. E. Hollyer.
ISBN 0 11 8841165 *not yet priced*

56 Procedure for the assessment of the conglomerate resources of the Sherwood Sandstone Group. D. P. Piper and P. J. Rogers.
ISBN 0 11 884143 2 £1.25

57 The conglomerate resources of the Sherwood Sandstone Group of the country around Cheadle, Staffordshire: Resource sheet SK 04. P. J. Rogers, D. P. Piper and T. J. Charsley.
ISBN 0 11 884144 0 *not yet priced*

58 The sand and gravel resources of the country west of Peterhead, Grampian Region: Resource sheet NK 04 and parts of NJ 94 and 95, NK 05, 14 and 15. A. A. McMillan and A. M. Aitken.
ISBN 0 11 884145 9 *not yet priced*

59 The sand and gravel resources of the country around Newbury, Berkshire: Resource sheets SU 46 and 57, parts of SU 36, 37 and 47. J. R. Gozzard.
ISBN 0 11 884146 7 *not yet priced*

60 The sand and gravel resources of the country south-west of Peterborough, in Cambridgeshire and East Northamptonshire: Resource sheets TL 09 and 19, and SP 98 and TL 08. A. M. Harrison.
ISBN 0 11 884147 5 *not yet priced*

61 The sand and gravel resources of the country north of Wrexham, Clwyd: Resource sheet SJ 35 and part of SJ 25. P. N. Dunkley.
ISBN 0 11 884148 3 *not yet priced*

62 The sand and gravel resources of the country around Dolphinton, Strathclyde Region, and West Luiton, Borders Region: Resource sheets NJ 04 and 14, and parts of NJ 05 and 15. A. A. McMillan, J. L. Laxton and A. J. Shaw.
ISBN 0 11 884149 1 *not yet priced*

63 The sand and gravel resources of the country of the valley of the Douglas Water, Strathclyde: Resource sheet NS 83 and parts of NS 82, 92 and 93. A. J. Shaw and E. P. F. Nickless.
ISBN 0 11 884150 5 *not yet priced*

64 The sand and gravel resources of the country between Wallingford and Goring, Oxfordshire: Resource sheet SU 68 and part SU 58. C. E. Corser.
ISBN 0 11 884151 3 *not yet priced*

65 The sand and gravel resources of the country around Hexham, Northumberland: Resource sheets NY 86 and 96. J. H. Lovell.
ISBN 0 11 884152 1 *not yet priced*

66 The sand and gravel resources of the country west of Chelmsford, Essex: Resource sheet TL 60. P. M. Hopson.
ISBN 0 11 884153 X *not yet priced*

Reports of the Institute of Geological Sciences

Other Reports

69/9 Sand and gravel resources of the inner Moray Firth. A. L. Harrison and J. D. Peacock.
ISBN 0 11 880106 6 35p

70/4 Sands and gravels of the southern counties of Scotland. G. A. Goodlet.
ISBN 0 11 880105 8 90p

72/8 The use and resources of moulding sand in Northern Ireland. R. A. Old.
ISBN 0 11 881594 0 30p

73/9 The superficial deposits of the Firth of Clyde and its sea lochs. C. E. Deegan, R. Kirby, I. Rae and R. Floyd.
ISBN 0 11 880617 3 95p

77/1 Sources of aggregate in Northern Ireland (2nd edition). I. B. Cameron.
ISBN 0 11 881279 3 70p

77/2 Sand and gravel resources of the Grampian Region. J. D. Peacock and others.
ISBN 0 11 881282 3 80p

77/5 Sand and gravel resources of the Fife Region. M. A. E. Browne.
ISBN 0 11 884004 5 60p

77/6 Sand and gravel resources of the Tayside Region. I. B. Paterson.
ISBN 0 11 884008 8 £1.40

77/8 Sand and gravel resources of the Strathclyde Region. I. B. Cameron and others.
ISBN 0 11 884028 2 £2.50

77/9 Sand and gravel resources of the Central Region, Scotland. M. A. E. Browne.
ISBN 0 11 8840169 £1.35

77/19 Sand and gravel resources of the Borders Region, Scotland. A. D. McAdam.
ISBN 0 11 884025 8 £1.00

77/22 Sand and gravel resources of the Dumfries and Galloway Region of Scotland. I. B. Cameron.
ISBN 0 11 884025 8 £1.20

78/1 Sand and gravels of the Lothian Region of Scotland. A. D. McAdam.
ISBN 0 11 884042 8 £1.00

78/8 Sand and gravel resources of the Highland Region. W. Mykura, D. L. Ross and F. May.
ISBN 0 11 884050 9 £3.00

Dd 696493 K8

Typeset for the Institute of Geological Sciences by
J. W. Arrowsmith Ltd, Winterstoke Road, Bristol

Printed in England for Her Majesty's Stationery Office by
Commercial Colour Press, London E7

INSTITUTE OF GEOLOGICAL SCIENCES

Natural Environment Research Council

Mineral Assessment Report 46

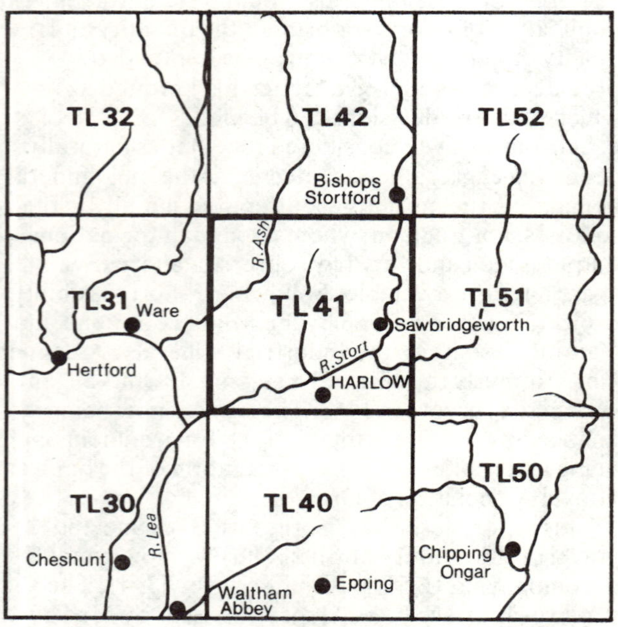

The sand and gravel resources of the country north of Harlow, Essex

Description of 1:25 000 resource sheet TL 41

P. M. H

ISBN 0 11 884107 6

London Her Majesty's Stationery Office 1979

The first twelve reports on the assessment of British sand and gravel resources appeared in the Report series of the Institute of Geological Sciences as a subseries. Report 13 and subsequent reports appear as Mineral Assessment Reports of the Institute.

Details of published reports appear at the end of this Report.

Any enquiries concerning this report may be addressed to Head, Industrial Minerals Assessment Unit, Institute of Geological Sciences, Keyworth, Nottingham NG12 5GG.

PREFACE

National resources of many industrial minerals may seem so large that stocktaking appears unnecessary, but the demand for minerals and for land for all purposes is intensifying and it has become increasingly clear in recent years that regional assessment of the resources of these minerals should be undertaken. The publication of information about the quantity and quality of deposits over large areas is intended to provide a comprehensive factual background against which planning decisions can be made.

Sand and gravel, considered together as naturally occurring aggregate, was selected as the bulk mineral demanding the most urgent attention, initially in the south-east of England, where about half the national output is won and very few sources of alternative aggregates are available. Following a short feasibility project, initiated in 1966 by the Ministry of Land and Natural Resources, the Industrial Minerals Assessment Unit (formerly the Mineral Assessment Unit) began systematic surveys in 1968. The work is now being financed by the Department of the Environment and is being undertaken with the cooperation of the Sand and Gravel Association of Great Britain.

This report describes the resources of sand and gravel of the country north of Harlow shown on the accompanying 1:25 000 resource sheet TL 41. The survey was by Mr P. M. Hopson assisted by Mr R. J. Marks and Mr J. W. Merritt. The work is based on a geological survey at the 1:10 560 scale in 1975–1976 by Dr B. Moorlock and Mr M. Heath of the East Anglia and South-Eastern England Unit. The inferred solid geology boundaries, based on the borehole data, are by Mr Hopson.

Mr J. W. Gardner, CBE, (Land Agent) was responsible for the negotiation of access to land for drilling. The ready cooperation of landowners and tenants in this work is appreciated. Information provided by members of the sand and gravel industry and by Hertfordshire and Essex County Councils is gratefully acknowledged.

Austin W. Woodland
Director

Institute of Geological Sciences
Exhibition Road
South Kensington
London SW7 2DE

12 March 1979

CONTENTS

FIGURES

MAP

TABLES

The sand and gravel resources of the country north of Harlow, Essex

Description of 1:25 000 resource sheet TL 41

P. M. Hopson

SUMMARY

The geological maps of the Institute of Geological Sciences, pre-existing borehole information and 98 boreholes drilled for the Industrial Minerals Assessment Unit, form the basis of the assessment of sand and gravel in the area north of Harlow, Essex.

All deposits in the area which might be potentially workable for sand and gravel have been investigated and a simple statistical method has been used to estimate the volume. The reliability of the volume estimates is given at the 95 per cent probability level.

The 1:25 000 map is divided into five resource blocks containing between 8.3 and 14.2 km² of sand and gravel. For each block the geology of the deposits is described and the mineral-bearing area, the mean thickness of overburden and mineral, and the mean gradings are stated. Detailed borehole data are also given. The geology, the position of the boreholes and the outlines of the resource blocks are shown on the accompanying map.

Bibliographic reference

Hopson, P. M. 1979. The sand and gravel resources of the country north of Harlow, Essex: description of the 1:25 000 resource sheet TL 41. *Miner. Assess. Rep. Inst. Geol. Sci.*, No. 46.

Note

All grid references quoted in this report fall within the 100-km grid square TL.

Author

P. M. Hopson, B.Sc
Institute of Geological Sciences
Keyworth, Nottingham NG12 5GG

INTRODUCTION

The survey is concerned with the estimation of resources, which include deposits that are not currently exploitable but have a foreseeable use, rather than reserves, which can only be assessed in the light of current, locally prevailing, economic considerations. Clearly, both the economic and the social factors used to decide whether a deposit may be workable in the future cannot be predicted; they are likely to change with time. Deposits not currently economically workable may be exploited as demand increases, as higher grade or alternative materials become scarce, or as improved processing techniques are applied to them. The improved knowledge of the main physical properties of the resource and their variability, which this survey seeks to provide, will add significantly to the factual background against which planning policies can be decided (Archer, 1969; Thurrell, 1971; Harris and others, 1974).

The survey provides information at the 'indicated' level 'for which tonnage and grade are computed partly from specific measurements, samples or production data and partly from projection for a reasonable distance on geological evidence. The sites available for inspection, measurement, and sampling are too widely spaced to permit the mineral bodies to be outlined completely or the grade established throughout' (Bureau of Mines and Geological Survey, 1948, p. 15).

It follows that the whereabouts of reserves must still be established and their size and quality proved by the customary detailed exploration and evaluation undertaken by the industry. However, the information provided by the survey should assist in the selection of the best targets for such further work. The following arbitrary physical criteria have been adopted:

a The deposit should average at least 1 m in thickness.
b The ratio of overburden to sand and gravel should be no more than 3:1.
c The proportion of fines (particles passing the No. 240 mesh BS sieve, about $\frac{1}{16}$ mm) should not exceed 40 per cent.
d The deposit must lie within 25 m of the surface, this being taken as the likely maximum working depth under most circumstances. It follows from the second criterion that boreholes are drilled no deeper than 18 m if no sand and gravel has been proved.

A deposit of sand and gravel that broadly meets these criteria is regarded as 'potentially workable' and is described and assessed as 'mineral' in this report. As the assessment is at the indicated level, parts of such a deposit may not satisfy all the criteria.

For the particular needs of assessing sand and gravel resources, a grain-size classification based on the geometric scale $\frac{1}{16}$ mm, $\frac{1}{4}$ mm, 1 mm, 4 mm, 16 mm has been adopted. The boundaries between fines (that is,

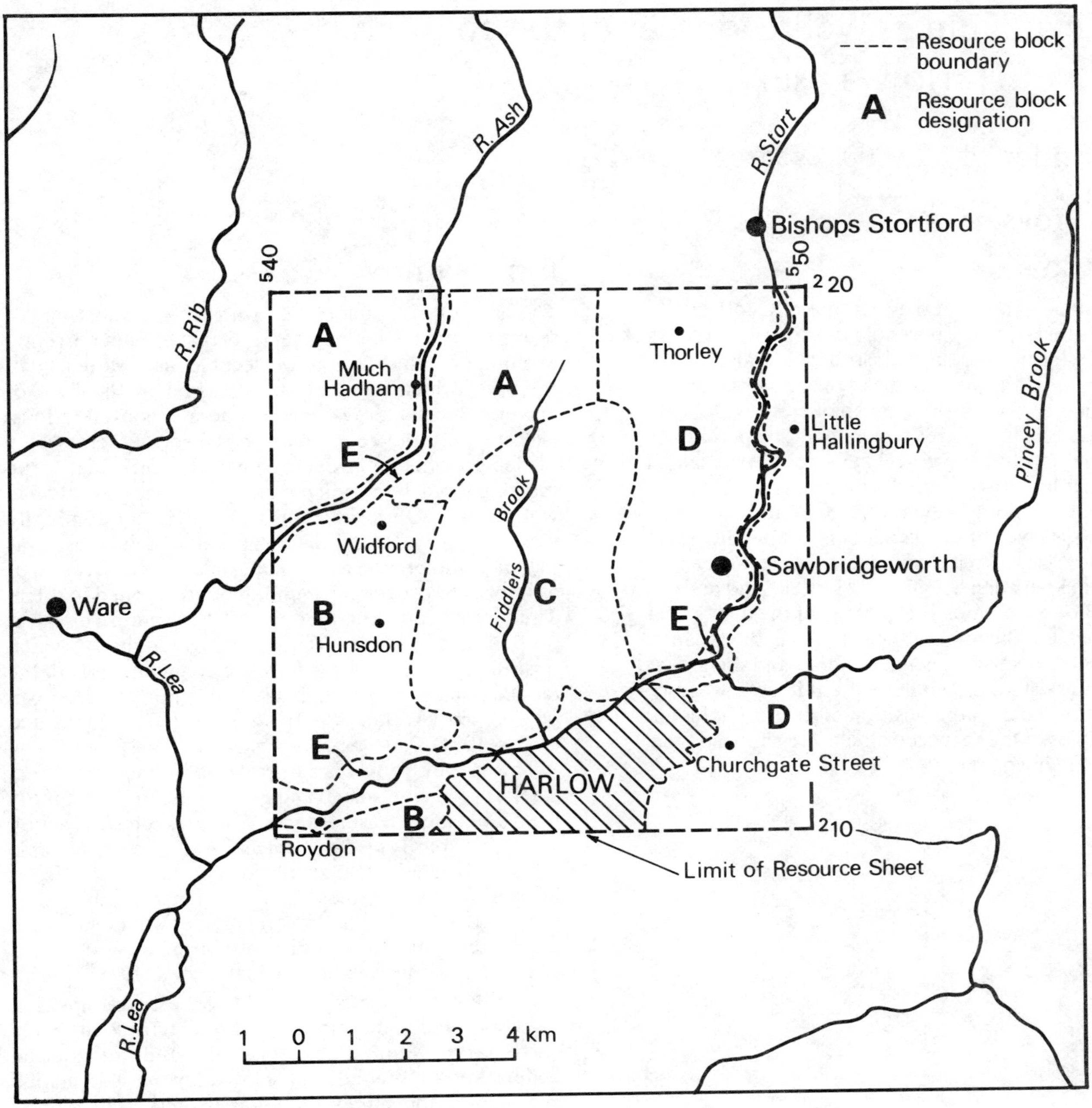

Figure 1 Sketch-map showing the location of sheet TL 41 and the position of the resource block boundaries

the clay and silt fractions) and sand, and between sand and gravel grade material, are placed at $\frac{1}{16}$ mm and 4 mm respectively (see Appendix C).

The volume and other characteristics are assessed within resource blocks, each of which, ideally, contains approximately 10 km² of sand and gravel. No account is taken of any factors, for example, roads, villages and high agricultural or landscape value, which might stand in the way of sand and gravel being exploited, although towns are excluded. The estimated total volume therefore bears no simple relationship to the amount that could be extracted in practice.

It must be emphasised that the assessment applies to the resource block as a whole. Valid conclusions cannot be drawn about the mineral in parts of a block, except in the immediate vicinity of the actual sample points.

DESCRIPTION OF RESOURCE SHEET

GENERAL

The resource sheet (TL 41) covers 100 km² of country north of Harlow, Essex, (see Figure 1) of which 57.6 km² is gravel-bearing. Harlow is 25 km west of Chelmsford, Essex, and some 34 km north-east of central London. Both glacial and fluvial deposits have been assessed, the latter, associated with the rivers Ash and Stort, being found only within block E.

No assessment has been made of the northern built-up area of Harlow New Town, lying within the demarcated planning boundary of 1981.

TOPOGRAPHY

The area consists of a plateau of Boulder Clay which generally lies between 76 m (250 ft) and 91 m (300 ft) Ordnance Datum and which slopes gently to the south. It has been dissected by the River Ash in the north-west and by the River Stort (see Figure 2). Exceptionally, the plateau rises above 91 m (300 ft) Ordnance Datum in three areas; at Jobbers Wood [457 196] and Mathams

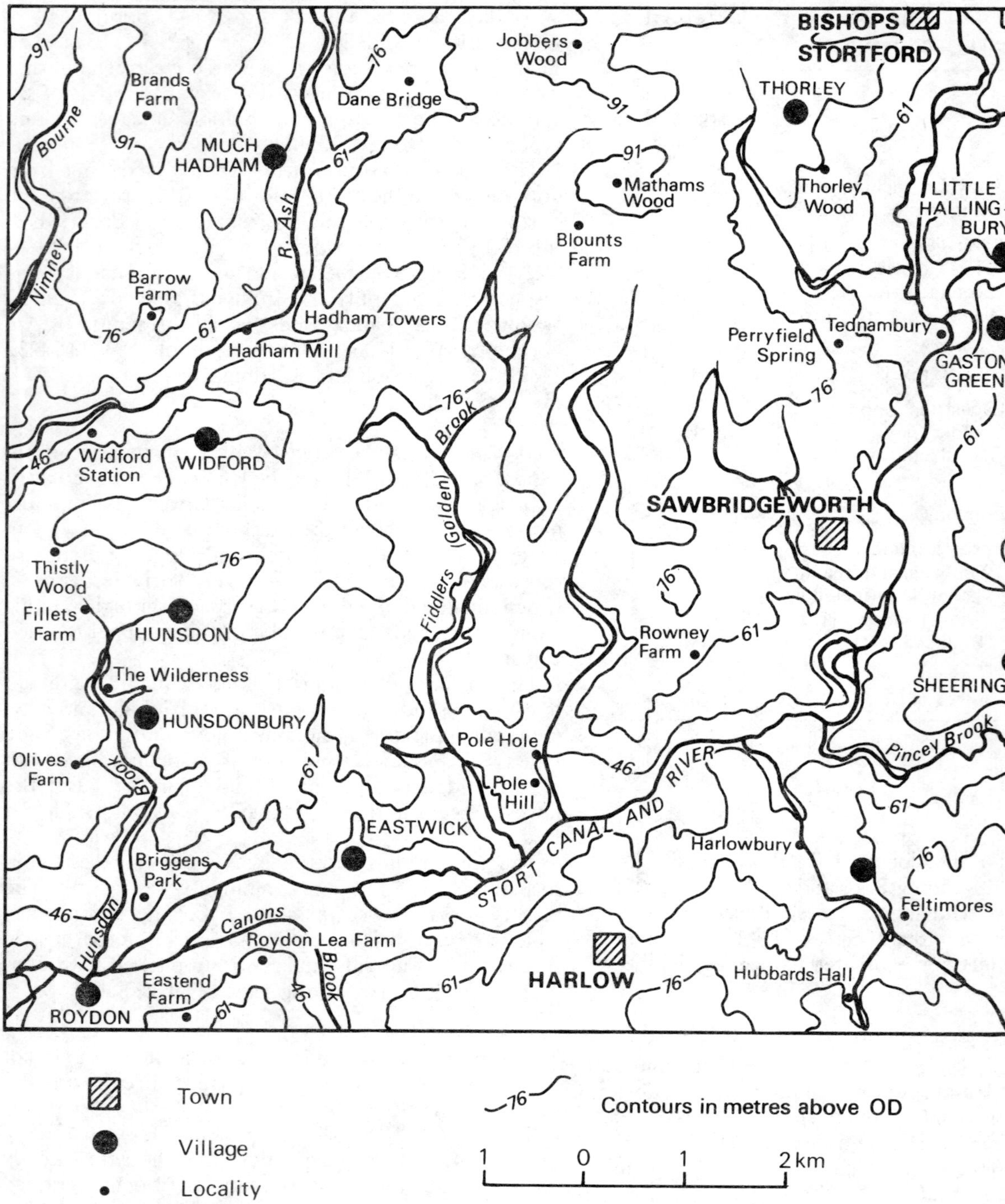

Figure 2 The topography in the area of sheet TL 41 showing the locations mentioned in the text

Wood [460 185] near Thorley, and at Brands Farm [413 191], Much Hadham, where the maximum elevation of 98 m (320 ft) Ordnance Datum is reached.

The south-westward flowing River Ash falls from about 60 m (197 ft) Ordnance Datum at Much Hadham to 38 m (125 ft) Ordnance Datum south-west of Widford, a distance of 6 km. This stream flows in a steep-sided valley in the area of Widford Station [408 158]. It is joined by a number of unnamed tributaries around Much Hadham and Widford. The deeply incised valley of the Nimney Bourne, whose headwaters drain much of the Boulder Clay country westward of Much Hadham, joins the Ash just beyond the western boundary of the sheet.

The largest river, the River Stort, earlier in its history flowed from Bishops Stortford to Harlowbury [480 120] in a deeply incised tunnel valley (See Figure 3 and Woodland, A. W., 1970), which was subsequently infilled with thick glacial deposits. However, it now flows in a broad valley over a veneer of its own fluvial deposits. The river, much of which was canalised in the late 1800's falls from 60 m (197 ft) Ordnance Datum at Bishops Stortford to about 30 m (99 ft) Ordnance Datum at Roydon, a distance of 17 km. Small tributaries join it on both banks, notably the Pincey, Fiddlers (Golden) and Canons brooks, which flow into the main stream north-east, north and west of Harlow respectively (Figure 2).

GEOLOGY

The assessment of resources is based on geological mapping at the six-inch scale by Dr B. Moorlock and Mr M. Heath in 1975–1976. The original Old Series one-inch geological survey by Messrs W. Whitaker, W. H. Penning, W. H. Dalton and F. J. Bennett was published in 1881 and the Drift edition in 1884. A geological classification of the deposits mapped is given in Table 1.

Table 1 Classification of mapped deposits; thicknesses quoted in metres

		Maximum recorded thickness m
DRIFT		
Recent and Pleistocene	Alluvium (including underlying terrace gravels)	8.1
	River Terrace Deposits	
	First Terrace	5.0*
	Second Terrace	1.3
	Undifferentiated Terrace	2.5*
	Head	3.9
	Head Gravel	6.3
	Lacustrine Deposits	15.2+†
	Boulder Clay	27.1
	Glacial Sand and Gravel	15.1
SOLID		
Eocene	London Clay	20.0
	Lower London Tertiaries	
	Woolwich and Reading Beds	17.0
	Thanet Beds (including Thanet Sand and the Bullhead Bed)	12.8
Upper Cretaceous	Upper Chalk	60.0

* The thickness of these deposits has been estimated as no borehole information is available.
† The base was not reached.

SOLID

The higher zones of the Upper Chalk, the Thanet Beds, the Woolwich and Reading Beds and the lowest part of the London Clay, with its associated Basement Bed, are represented in the area of Sheet TL 41 (see Figure 4). The low regional dip is estimated to be a few degrees to the south-east.

Upper Chalk:
The Upper Chalk is a soft white well-bedded and jointed limestone with beds and courses of nodular and thin tabular flint. Its thickness in the area of sheet TL 41 is of the order of 60 m (see non-IMAU borehole record TL 41 NW 43 [4196 1662]). Upper Chalk is only exposed in the north-west quadrant of the sheet but it subcrops under Pleistocene and Recent deposits in the extreme south-west and in the deep, drift-filled, Stort channel from Bishops Stortford to Harlowbury.

Numerous small exposures of Upper Chalk occur in the area around Much Hadham and Widford, and also in the deeply incised Nimney Bourne at Nobland Green [401 176].

Lower London Tertiaries:
The Lower London Tertiaries comprise two formations: the Thanet Beds, and the Woolwich and Reading Beds. Summaries of lithologies present on sheet TL 41 are given below but more detailed work has been published by Prestwich (1850, 1852, 1854) and other work comprehensively reviewed by Hester (1965).

Thanet Beds: This formation rests disconformably on the eroded surface of the Chalk. The base is marked by a well-developed flint pebble bed, with glauconitic quartz sand matrix (the Bullhead Bed). It reaches a maximum recorded thickness in IMAU boreholes SW 17 [4072 1065] and SW 22 [4135 1062] of 0.5 m. The Thanet Sand above comprises drab olive green fine and medium glauconitic sands with varying amounts of silt and clay. The sands are commonly micaceous, and sometimes burrowed and finely bedded.

There is no exposure of the Thanet Beds, but from borehole data they are known to thin towards the north-west. Borehole records from Rowney Farm (SE 78 [4697 1372]) and north of Widford (NW 44 [4222 1658]) proved 8.2 m and 4.9 m of Thanet Beds respectively, but only the Bullhead Bed, overlain by Woolwich and Reading Beds clays, is present in IMAU borehole NW 4 [4068 1683]. In borehole NW 11 [4232 1989] the Woolwich and Reading Beds rest directly on the Upper Chalk.

Woolwich and Reading Beds: With the exception of an exposure [4305 1776] of clayey micaceous sand in a drainage ditch north of Hadham Towers Pit, this formation is concealed beneath drift deposits. From borehole data, the Woolwich and Reading Beds are represented by up to 17.0 m of very variable deposits, which may be divided into two broad lithological units: a lower sand and clayey sand unit and an upper mottled clay unit.

The lower unit comprises fine to medium-grained micaceous sand with varying amounts of silt and clay. Generally pale green, the unit can be mottled in a wide variety of colours. Up to three flint pebble beds with well rounded pebbles are found near the base; the lowest, except in the extreme NW, rests on the Thanet Beds.

The upper unit of mottled, waxy, stiff clays rests conformably on the lower predominantly sandy unit. The clay is generally brown mottled red and green but a wide variety of bright colours have been noted. It is in places sandy and silty and is rarely packed with concretionary carbonate nodules and bivalve shells.

London Clay
The London Clay which forms rock-head over much of the sheet rests without apparent break on the Woolwich and Reading Beds.

The lowest beds of the London Clay, commonly between 4.0 m and 5.0 m thick, are characteristically dark greyish green, clayey, silty, pyritic fine-grained sands and are called the Basement Bed. A pebble bed of black and brown well-rounded flints marks the base. It was formerly exposed at the Gaston Green Pit [4950 1625] and the Hadham Towers Pit [4310 1746].

Higher beds of the London Clay are firm, often fissured, bluish grey and dark grey clays, which often contain selenite crystals and pyrite nodules. When exposed as at Eastend Farm [419 101] and in Fiddlers (Golden) Brook [445 140], the London Clay is weathered to brown mottled yellowish brown silty clay.

A fossiliferous, waxy reddish brown clay, approximately 5.0 m in thickness, occurs in the valley from Feltimores [490 112] to Hubbards Hall [484 104]. It is probable that the maximum thickness (20.0 m+) of London Clay on this sheet occurs in this area.

DRIFT

Glacial Sand and Gravel
The deposits of Glacial Sand and Gravel represent the main resource of aggregate in the area of sheet TL 41.

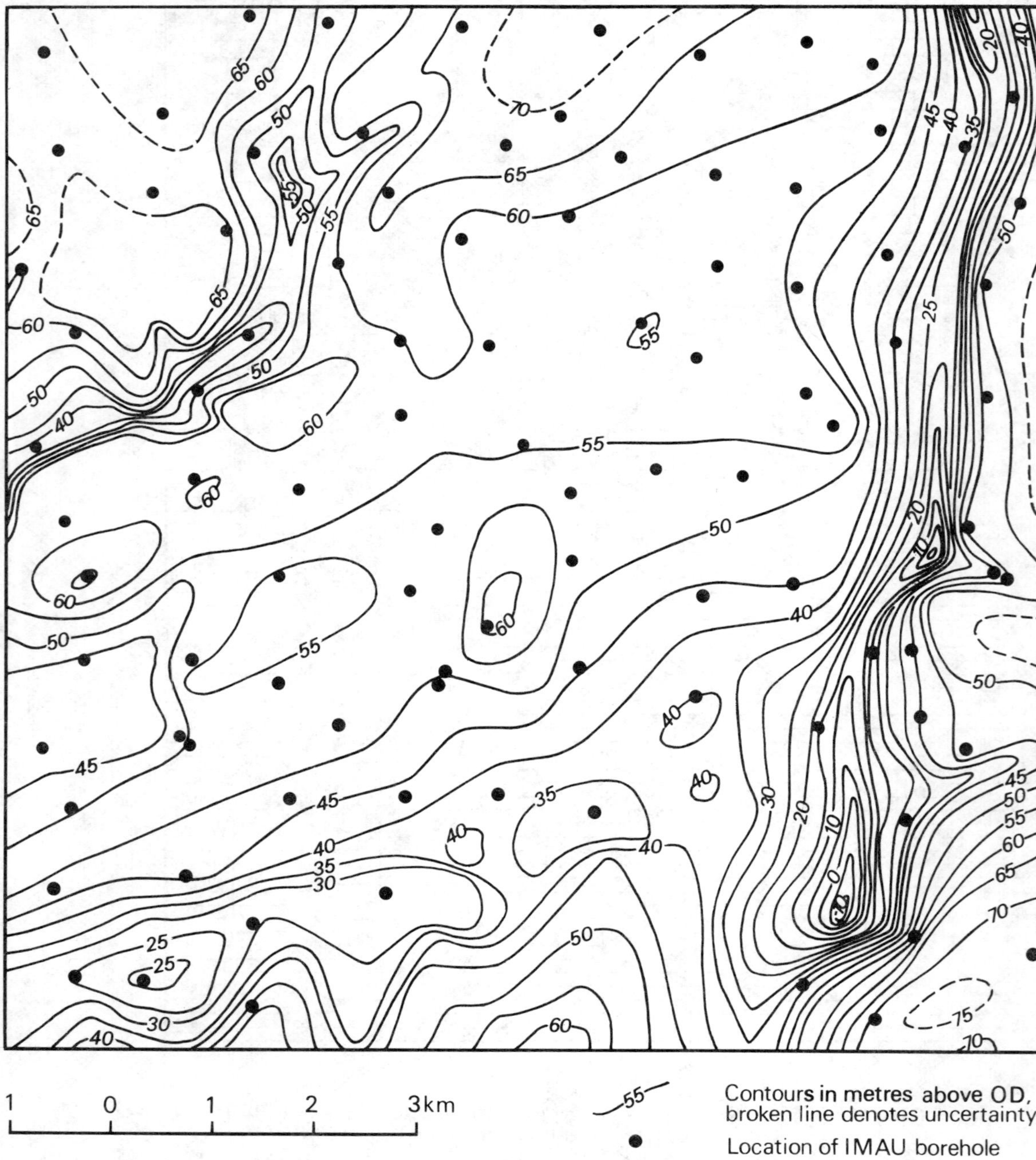

Figure 3 Contours on the sub-drift surface

They are now, however, only being worked at the Gaston Green Pit [495 164]. Aggregate has also been won, in the past, from pits at Hadham Towers [430 170] and Pole Hill [452 123, 458 128] and to a lesser extent from a pit [442 197] north-east of Dane Bridge.

The mineral is heterogeneous, ranging from sand to sandy gravel and gravel, generally 'clayey' or 'very clayey' (see Figure 11), and is often ironstained. Where ironstaining is heavy, as in the west and south-west of the sheet area, the deposit becomes locally cemented into a hard ferrocrete rock. This rock occurs particularly where the deposit is classified as sandy gravel and gravel.

Beds of silt, sometimes laminated and cross-bedded, occur locally in the sand and gravel as in IMAU borehole SW 21 [4178 1165] and in the pits at Hadham Towers and Gaston Green. At Hadham Towers a bed of finely laminated silt (see Figure 5) with chalky sand seams is intercalated between two beds of sandy gravel.

The pit north-east of Dane Bridge shows 4 m of 'clayey' sandy gravel resting on chalky Boulder Clay, and as with the sandy gravels at Hadham Towers, many derived fossil fragments are present.

The only other current exposure is at Gaston Green Pit where up to 5.5 m of sand and sandy gravel, cross-bedded on a large scale with units up to 1 m in thickness, are seen. In the north-east corner of this exposure 4 m of bedded gravel and sand, overlain by a structureless 'very clayey' gravel, rest with a sharp junction on laminated chalky silts similar to those of Hadham Towers Pit.

The origin of seams of chalky Boulder Clay found within the Glacial Sand and Gravel in many of the boreholes is conjectural; they are probably flow tills (Boulton, 1972).

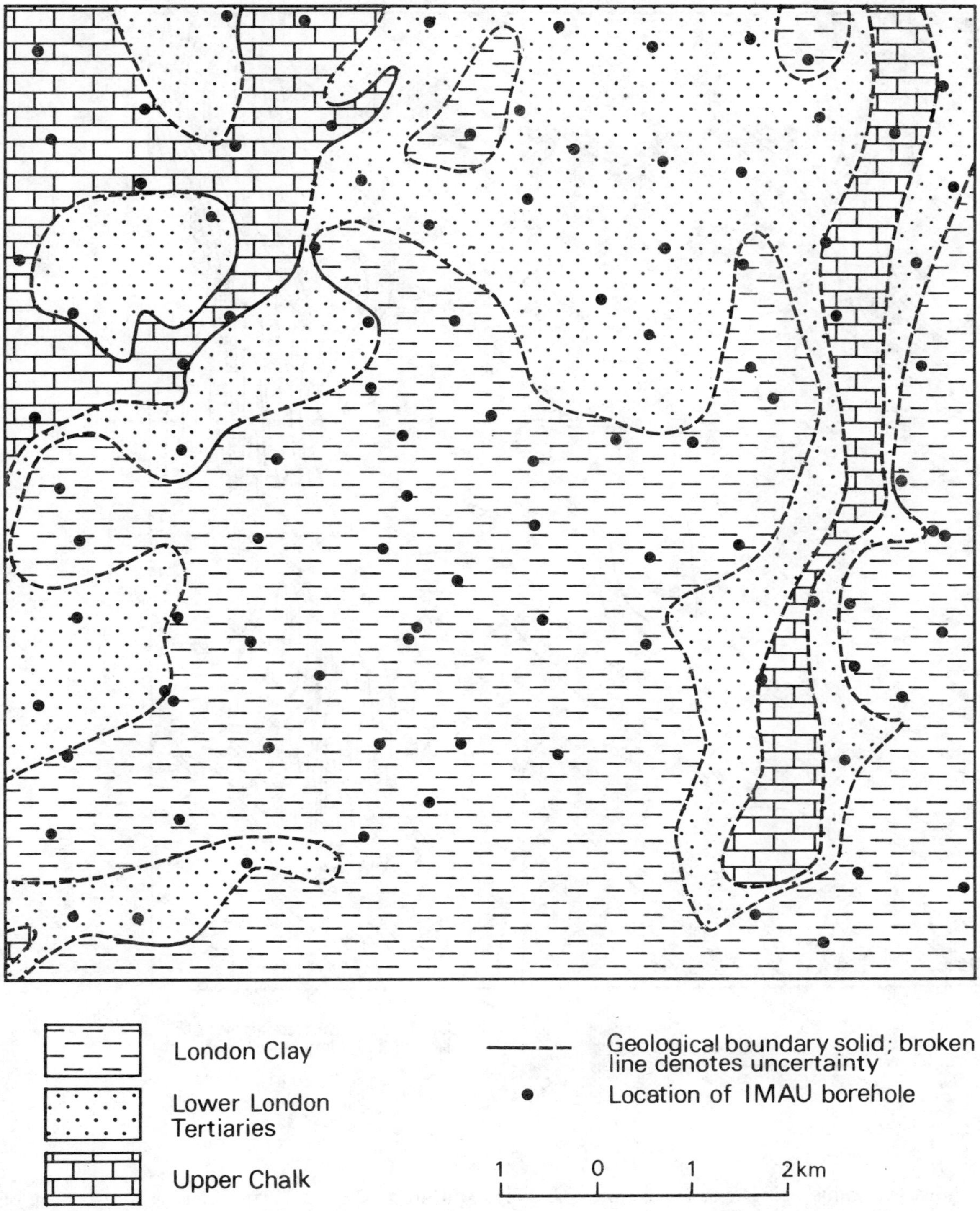

Figure 4 Sketch-map showing the sub-drift geology of sheet TL 41 incorporating new information from IMAU boreholes

Boulder Clay

The Boulder Clay is usually a grey to bluish grey stiff chalky clay with some silt and rarely sand. It contains many pebbles predominantly of rounded to angular flint, and rounded quartz and quartzite with scattered clasts of varying angularity and shape, of schist, limestone, mudstone and sandstone. Rarely, a bluish black very stiff clay with sand-sized chalk pellets is developed near the base of the Boulder Clay (e.g. IMAU borehole SW 25 [4272 1236]); it is often packed with quartz pebbles and fossil debris.

On weathering, the bulk of the Boulder Clay becomes ochre brown in colour but grey mottling persists around unleached chalk pellets. Weathering from the surface is generally up to 3 m deep but exceptionally it is greater, as in IMAU borehole NE 22 [4663 1660], where it was proved to 7.9 m.

Occasionally the Boulder Clay becomes very silty and sandy and lenses of chalk gravel occur in some boreholes (e.g. NE 16 [4539 1795]). These gravels are usually saturated with water under sub-artesian head.

In boreholes NE 23 [4627 1552] and SW25 [4272 1236], rafts of Chalk and London Clay respectively were encountered.

The Boulder Clay appears to cut down into and in some cases through (e.g. NE 32 [4848 1760]) the underlying Glacial Sand and Gravel.

Clayton (1957) considered the Boulder Clay above the Glacial Sand and Gravel to be the equivalent of the Springfield Till. The Boulder Clay underlying the Glacial Sand and Gravel, and termed by Clayton the Maldon Till, is impersistent and proved only in a few boreholes on the western margin of this sheet.

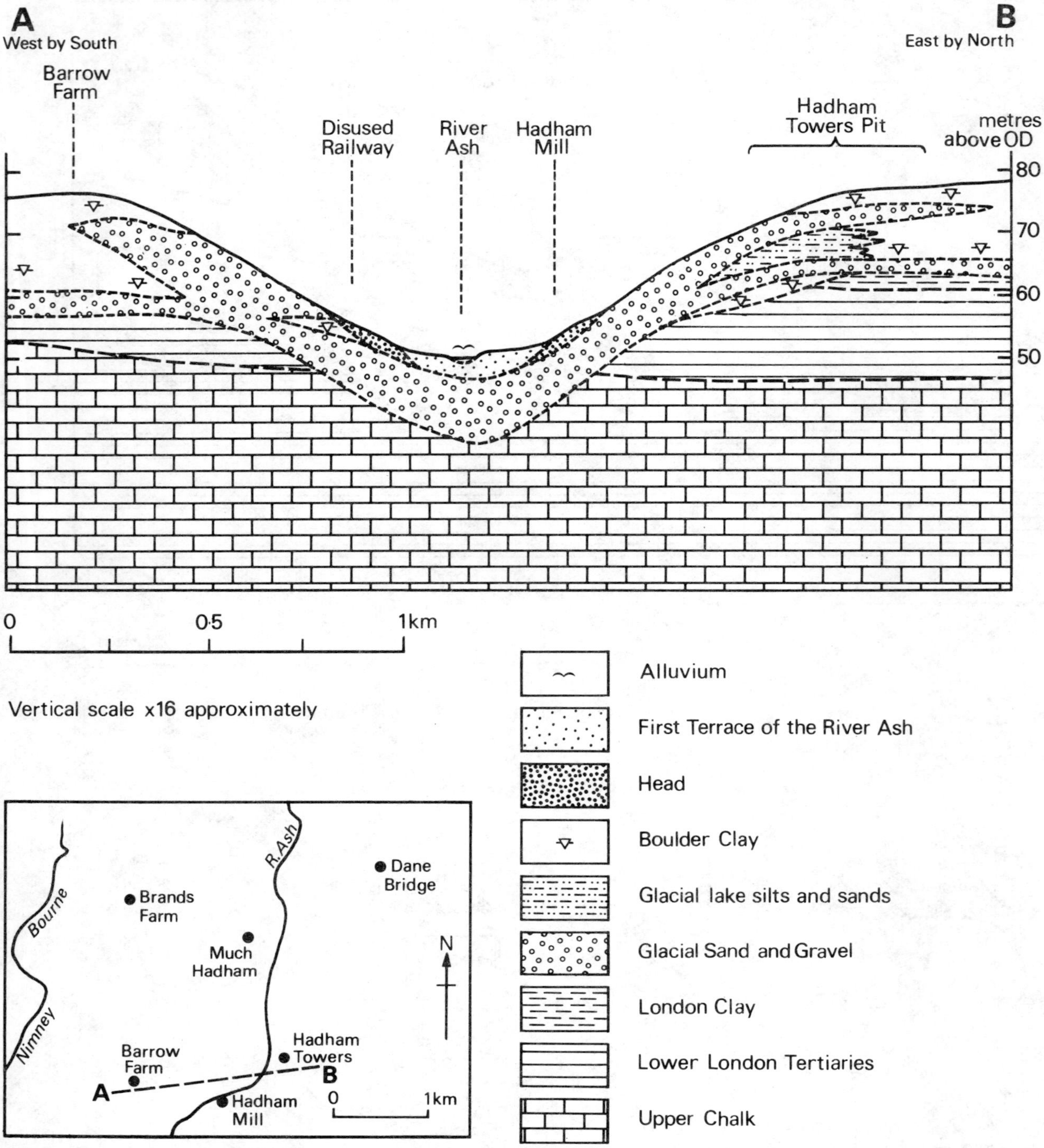

Figure 5 Sketch-diagram to demonstrate the probable relationship of the drift deposits in the vicinity of Barrow Farm to those at Hadham Towers

Lacustrine Deposits

The only exposure of these deposits is in a ditch eastward of Perryfield Spring [485 168] but they have also been proved in the nearby IMAU borehole NE 33 [4856 1676]. The ditch section shows dark brown to pale grey silty laminated clay and fine shelly sand with beds of thin peat containing wood, rootlets, mosses, seeds and bones.

Head Gravel

These are poorly sorted, and apparently structureless, clayey, angular to rounded gravels with varying amounts of sand, which overlie both the Boulder Clay and Glacial Sand and Gravel. The only small exposure of this material, at the junction [498 187] of Port Lane and the B1005 near Little Hallingbury, shows 1.5 m of clayey fine angular flint gravel resting on decalcified Boulder Clay. The maximum recorded thickness, 6.3 m, is in IMAU borehole NE 36 [4977 1813] where 2 m of 'clayey' sandy gravel rests on 2 m of sand and silt and this on 2.3 m of 'clayey' pebbly sand. These gravels occur in patches around Little Hallingbury [500 176], Hunsdon [416 141] and Churchgate Street [485 115]. Their origin is not clear but their structureless nature and poor sorting at outcrop suggest that they may have been deposited during the melting of dead ice.

Head

Head of dominantly sandy silt is well developed in the valleys of the rivers Stort and Ash. It is however variable in composition being predominantly gravelly when overlying Glacial Sand and Gravel and clayey when overlying either Boulder Clay or London Clay. In

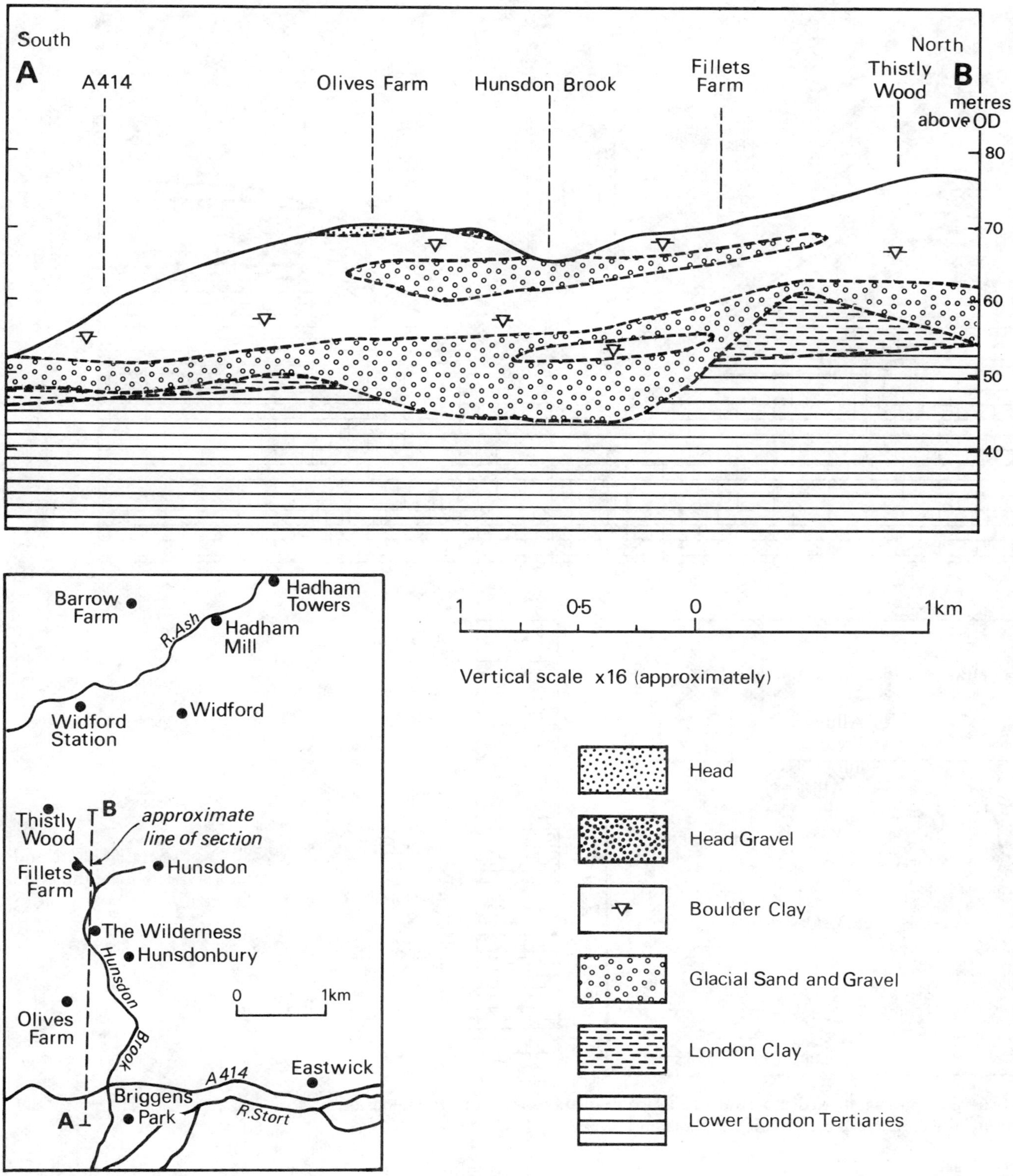

Figure 6 Sketch-diagram from Thistly Wood to the A414 demonstrating the relationship between the drift deposits

some areas, as at Tednambury [491 169] and Eastend Farm [419 101], very clayey Head is directly derived from the London Clay.

Numerous patches of Head have been mapped at high elevations on the Boulder Clay plateau, principally at Jobbers Wood [457 196], Blounts Farm [460 175] and Olives Farm [405 126]. IMAU boreholes and IGS trial pits have shown them to be composed of occasionally laminated brown silts, and silty clays.

River Terrace Deposits

Terraces are found along the valleys of the River Stort and River Ash where the First and Second Terraces form low features, 1 to 2 m and 3 to 4.5 m, respectively, above the flood plain.

The Undifferentiated Terrace [429 181] mapped at Much Hadham is an erosional feature 2 to 3 m above the flood plain and may be a remnant, left by the meandering River Ash, of a once continuous sheet of Glacial Sand and Gravel.

The First and Second terraces are typically gravelly and chalky and usually clayey and silty in the topmost metre. In the valley of the River Stort the First Terrace has been worked [495 195] at Bishops Stortford, the Second Terrace [410 103] at Roydon and both at Sheering [485 136]: the First Terrace of the River Ash has been dug [423 168] at Widford.

Table 2 Mean percentage composition by weight of mineral, by block and type of deposit

Block	Chalk	Flint and chert	Quartz and quartzite	Fossil debris	Sandstone	Others*
A	6	61.25	22	0.25	6	4.5
B	13	62	14	1	6	4
C	2	71	17	–	6	4
D	5	63.25	18.25	0.5	8	5
E	10.25	72	9	0.5	4.25	4
A to E	7	66	16.5	0.5	6	4
Deposit type						
Glacial Sand and Gravel	7.5	64	17	0.5	7	4
Alluvium†	3.5	79	11	–	3.5	3
First and Second Terrace	—	86.5	8	–	3	2.5
Head Gravel	1	81	8	–	6	4

* Including limestone, igneous, metamorphic, etc.
† Including underlying terrace gravel

Alluvium (including underlying terrace gravels)
In general, the Alluvium is represented by silty sandy clay with Recent gastropod shells and seams of chalky sand resting on variable thicknesses of peat and chalky, flinty terrace gravel.

Large spreads of Alluvium occur in the River Stort valley but in the valley of the River Ash the Alluvium forms a narrow sinuous belt closely following the present stream. The Alluvium is generally thin but, where the underlying terrace gravels are thick, up to 8.1 m of deposits have been proved (e.g. SW 17 [4072 1065]).

COMPOSITION OF SAND AND GRAVEL DEPOSITS
Deposits of Pleistocene and Recent age cover 98 per cent of the resource sheet. Of the nine formations mapped, six contain potentially workable sand and gravel: Glacial Sand and Gravel, Head Gravel, three terraces of different ages and the terrace gravels underlying the Alluvium.

The potentially workable sand and gravel deposits are predominantly composed of flint with varying amounts of chalk, quartz, quartzite, sandstone and traces of mudstone, limestone, metamorphic and igneous rocks. The Glacial Sand and Gravel has supplied much of the material from which the other gravels are derived. Although upon reworking the soft deleterious (for aggregates), less durable, material is usually destroyed, the sand and gravel underlying the Alluvium contains some chalk.

Table 2 shows the percentage composition, by weight, of mineral by block and type of deposit, based on a count of all samples within the range 4 mm to 16 mm. Of the five blocks, block E shows the greatest deviation from the mean, having a higher content of flint and less sandstone, quartz and quartzite than any other block.

Figure 7 shows the grading characteristics of mineral in IMAU boreholes in the area of sheet TL 41.

Glacial Sand and Gravel: All twelve IMAU mineral categories have been proved within the Glacial Sand and Gravel, demonstrating the extreme variability of this deposit on sheet TL 41. However, the categories gravel, sandy gravel and 'clayey' sandy gravel represent 62 per cent of the proved mineral while sand, 'clayey' sand and 'very clayey' sand represent only 8 per cent.

The gravel fraction is composed of angular to subangular with some rounded, black, brown and white flint, with rounded to well rounded white, pale brown and occasionally red quartz, and purple, grey or liver-coloured quartzite. Some pebbles of rounded white and pale yellow chalk, and subrounded to tabular sandstone occur. A trace of fossil debris occurs in 50 per cent of the IMAU boreholes usually in the form of fragmentary belemnite guards and thick-shelled bivalves. Mudstone, limestone and various igneous and metamorphic rocks were represented by a few pebbles in most of the samples.

The sands are predominantly medium-grained with fine and some coarse grades. Generally the sand is subangular to rounded quartz with some angular flint and a trace of chalk and sandstone. However, the coarse sand, when present in significant quantities, is composed of equal proportions of angular flint and subrounded to rounded quartz.

Some ironstaining and cementing does occur but is only pronounced in the Hunsdon and Widford areas near the western margin of the sheet (see for example: IMAU borehole NW 2 [4050 1862]).

Occasionally gravels with a very high chalk content (for example 68 per cent in IMAU borehole SW 23 [4262 1449]) are found but they usually occur in association with the Boulder Clay and not the Glacial Sand and Gravel. They are considered to be proximal outwash gravel derived from the chalky till.

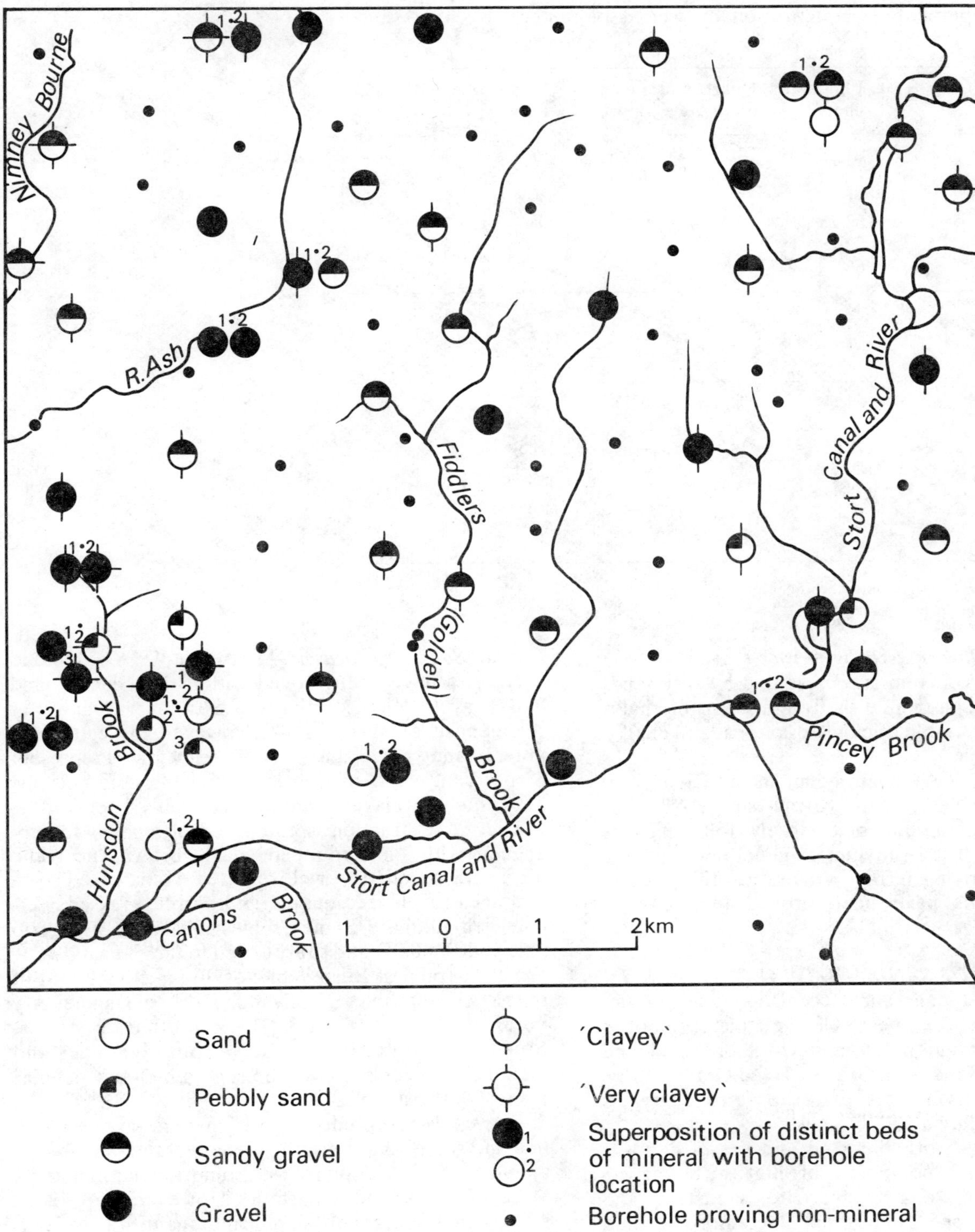

Figure 7 Grading characteristics of sand and gravel (mineral) proved in IMAU boreholes, expressed as the mean grading of each distinct bed of mineral

Head Gravel: Only one borehole (NE 36 [4977 1813]) was drilled in undoubted Head Gravel. The mineral falls within the category 'very clayey' sandy gravel, and is predominantly of flint with some quartz, quartzite and sandstone, and a trace of chalk. The sand is predominantly medium-grained with fine and some coarse grade ironstained quartz, with some flint.

First and Second Terrace: These deposits have a very high flint content (> 85 per cent) with some quartz and quartzite, and a trace of sandstone. The sand is predominantly of rounded quartz with rounded to subangular flint and a trace of subangular sandstone. The two IMAU boreholes SE 42 [4847 1404] and NW 14 [4231 1679] proved 'clayey' gravel and gravel respectively.

Alluvium (including underlying terrace gravels). Five boreholes drilled in the lower reaches of the River Stort proved sand and gravel underlying a silty sandy clay and peat. The sand and gravel deposits proved all fall within the category gravel and have a composition intermediate between terrace and Glacial Sand and Gravel. They have a high pebble content (mean of 61 per cent) composed predominantly of flint (79 per cent) with some rounded quartz, quartzite and sandstone and a trace of rounded chalk.

The sand is medium and coarse grained with a trace of fine grade and it is composed of angular flint and rounded quartz in equal proportions with a trace of chalk and sandstone.

Table 3 The sand and gravel resources of sheet TL 41: summary of statistical results

	Area		Mean thickness		Volume of mineral			Mean grading percentage		
						Limits at the 95% confidence level				
Resource Block	Block	Mineral	Overburden	Mineral				Fines	Sand	Gravel
	km²	km²	m	m	m³×10⁶	±%	± m³×10⁶	$-\frac{1}{16}$ mm	+16–4 mm	+4 mm
A	19.6	11.5	7.1	6.3	72	24	17	12	46	42
B	14.7	12.0	6.9	6.8	82	35	29	14	51	35
C	19.1	11.6	9.5	5.6	65	20	13	8	49	43
D	29.8	14.2	6.7	6.5	92	19	18	11	54	35
E	8.7	8.3	1.6	4.2	35	47	16	7	43	50
A to E	91.9	57.6	6.1	5.8	334	13	43			

THE MAP

The sand and gravel resource map is folded into the pocket at the end of this report. The base map is the Ordnance Survey 1:25 000 Outline Edition in grey, on which the topography is shown by contours in green, the geological data in black and the mineral resource information in shades of red.

Geological data: The geological boundary lines, symbols, etc., shown are taken from the geological maps of this area, which was surveyed at the scale of 1:10 560 in 1975–1976. Borehole data, which include the stratigraphic relation and mean particle-size distribution of the sand and gravel samples collected during the assessment survey, are also shown.

The geological boundaries represent the best available interpretation of the information available at the time of survey. However, it is inevitable, particularly with glacial deposits (such as those in the area of sheet TL 41) which change rapidly vertically and laterally, that local irregularities or discrepancies will be revealed by some boreholes (for example, in boreholes SE 39 and SW 27). These are taken into account in the assessment of resources (see below and Appendix B).

Mineral resource information: The mineral-bearing ground is subdivided into resource blocks (see Appendix A). Within a resource block the mineral is subdivided into areas where it is 'exposed' and areas where it is present beneath overburden. The mineral is identified as 'exposed' where the overburden, commonly consisting only of soil and subsoil, averages less than 1 m (3.25 ft) in thickness. Beneath overburden the mineral may be continuous (or almost continuous) or discontinuous. The recognition of these categories is dependent upon the importance attached to the proportion of boreholes that did not find potentially workable sand and gravel and the distribution of barren boreholes within the block. The mineral is described as 'almost continuous' if it is present in 75 per cent or more of the boreholes in a resource block. The 'discontinuous' category has not been recognised on the present sheet.

Areas where bedrock outcrops, where boreholes indicate the absence of sand and gravel beneath cover and where sand and gravel beneath cover is interpreted to be not potentially workable, are uncoloured on the map; where appropriate, the relevant criterion is noted. In such areas it has been assumed that mineral is absent except in infrequent and relatively minor patches which can neither be outlined nor assessed quantitatively in the context of this survey. The use of the term 'excessive overburden' on the map, implies that the overburden exceeds 18 m in thickness or that it is more than three times the thickness of the sand and gravel, or both. Areas of unassessed sand and gravel, for example, built-up areas, are indicated by a red stipple.

The area of the exposed sand and gravel is measured from the mapped geological boundary lines. The whole of this area is considered as mineral, although it may include small areas where sand and gravel is not present or is not potentially workable. Inferred boundaries have been inserted where sand and gravel beneath cover is interpreted to be not potentially workable or absent. Such boundaries (for which a distinctive symbol is used) are drawn primarily for the purpose of volume estimation. The symbol is intended to convey an approximate location within a likely zone of occurrence rather than to represent the breadth of the zone, its size being limited only by cartographic considerations. For the purpose of measuring areas the centre-line of the symbol is used.

RESULTS

The statistical results are summarised in Table 3. Fuller grading particulars are shown in Figure 8.

Accuracy of the results: For the five resource blocks (A to E) assessed statistically, the accuracy of the results at the symmetrical 95 per cent probability level (that is, it is probable that 19 times out of 20 the true volume lies within the given limits) varies between 19 and 47 per cent. However, the true values are more likely to be nearer the volumes calculated than either of the limits. Moreover, it is probable that roughly the same percentage limits would apply for the estimate of volume of a much smaller parcel of ground (say 100 hectares) containing similar sand and gravel deposits if the results from the same number of sample points (as provided by, say, ten boreholes) were used in the calculation. Thus, if closer limits are needed for the quotation of reserves of parts of a block, it can be expected that data from more than ten sample points will be required even if the area is quite small. This point can be illustrated by considering the whole of the statistically assessed sand and gravel on the sheet. The volume, 334 million m³, can be estimated to limits of

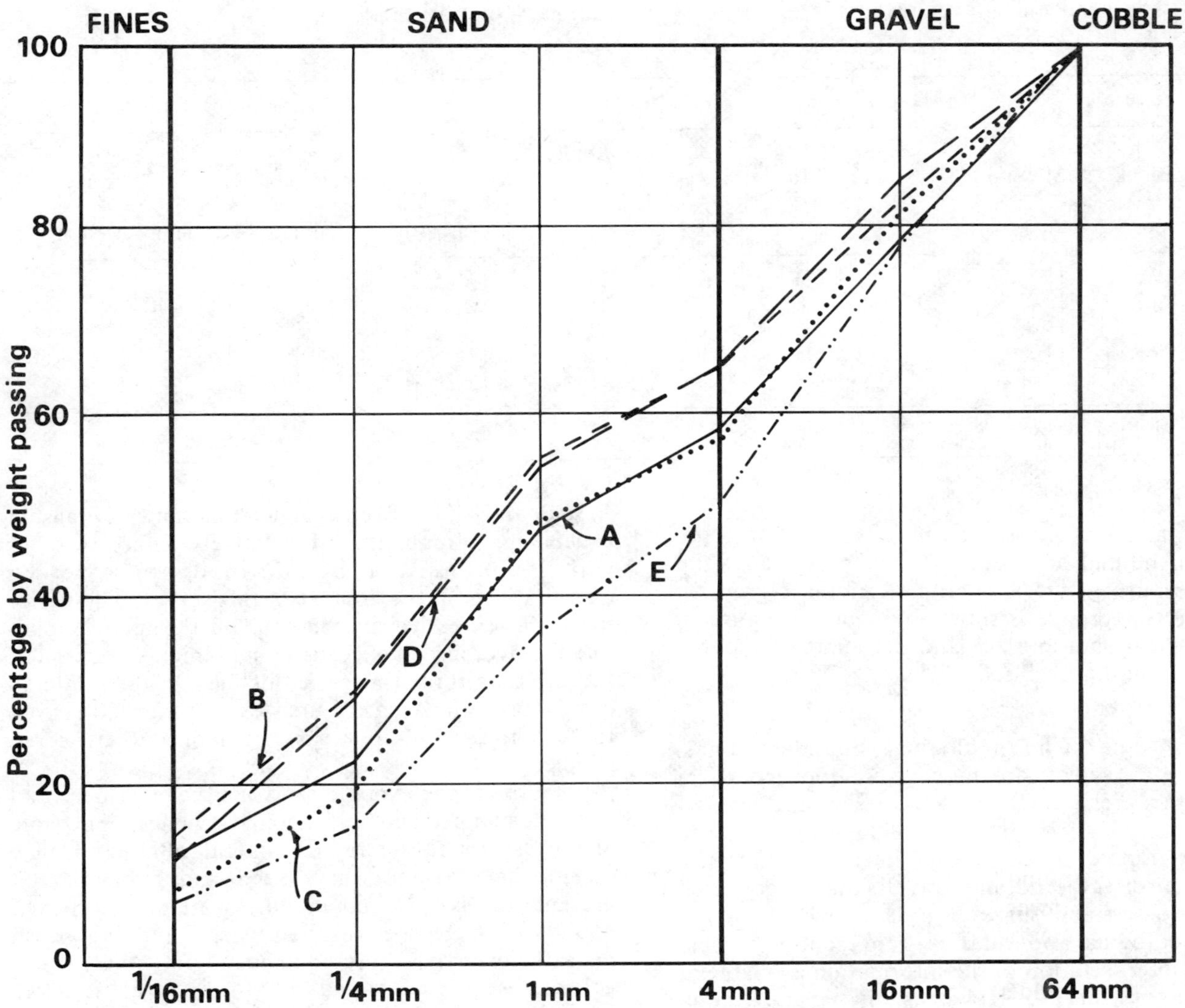

Block	Percentage by weight passing $\frac{1}{16}$ mm	$\frac{1}{4}$ mm	1 mm	4 mm	16 mm	64 mm
A	12	22	47	58	79	99
B	14	30	55	65	83	100
C	8	19	46	57	81	99
D	11	29	54	65	85	100
E	7	15	36	50	78	100

Figure 8 Particle-size distribution for the assessed thickness of mineral in resource blocks A to E

±13 per cent at the symmetrical 95 per cent probability level, by a calculation based on 61 data points in blocks A to E.

However, it must be emphasised that this quoted volume of sand and gravel bears no simple relationship to the amount that could be extracted in practice, as no allowance has been made in the calculations for any restraints (such as existing roads and buildings) on the use of land for mineral working.

NOTES ON RESOURCE BLOCKS

The block boundaries (see Figure 1) have been drawn to separate Glacial Sand and Gravel, both exposed and covered by Boulder Clay (blocks A, B, C, and D), from the fluvial deposits of the rivers Ash and Stort (block E).

Block A

This block comprises two areas of mineral separated by the fluvial deposits of the River Ash, which have been assessed in block E. The deposits are entirely glacial in origin with the superposition (normal for the sheet) of Boulder Clay upon Glacial Sand and Gravel.

The Boulder Clay overburden is thickest (19.7 m+, see, e.g. NE 15 [4281 1531]) on the relatively flat interfluves, where it appears to cut down into (e.g. NE 14 [4532 1891]) the Glacial Sand and Gravel. The overburden in boreholes used in the assessment ranges in thickness from 15.5 m in NW 18 [4370 1818] to 2.2 m in NW 22 [4436 1978] and has a mean of 7.1 m.

The Glacial Sand and Gravel, which represents the mineral in this block, is generally buried beneath Boulder Clay but is exposed on the sides of the Ash valley. It ranges in thickness from 9.9 m in NW 22

[4436 1978] to 3.3 m in NW 3 [4014 1741] and has a mean of 6.3 m. However field evidence and other borehole records show that up to 13.5 m of mineral is present in the area south-east and east of Barrow Farm [417 171].

There are at present two main exposures of mineral in this block, at Dane Bridge [442 197] and at Hadham Towers [430 170]. The two small pits north of Dane Bridge show gravelly topsoil grading down into 4.0 m of sandy gravel containing fossil debris. This sandy gravel rests on bluish grey mottled brown chalky flinty clay whose base was not seen.

The Hadham Towers Pit shows a more complicated stratigraphy (see Figure 5), with two glacial gravels, the uppermost with derived fossils, separated by laminated silts and chalky fine sands. These silts and sands interdigitate with a chalky Boulder Clay towards the east.

The deposits at Hadham Towers and the thick sands and gravels around Barrow Farm may be attributed to ponding in a blocked pre-glacial valley. This valley has subsequently been deepened by the River Ash.

The volume of mineral in block A is calculated to be 72 million m³, ±24 per cent (see Table 3) at the 95 per cent probability level. Its mean grading is fines 12 per cent, sand 46 per cent and gravel 42 per cent (including 1 per cent of cobble grade) giving an overall mineral classification of 'clayey' sandy gravel.

Block B

This block, as with block A, comprises two areas of mineral-bearing ground, which are, in this case, separated by the lower reaches of the River Stort whose associated mineral deposits have been assessed in block E.

As in block A, Boulder Clay usually rests on Glacial Sand and Gravel which is exposed on the valley sides. This sequence is demonstrated, for example, in IMAU boreholes NW 10 [4183 1543] and SW 16 [4048 1151]. However, in the area around Hunsdon and Hunsdonbury, in a shallow broad depression floored by Woolwich and Reading Beds, lenses of chalky, flinty brown clay and 'clayey' sandy gravel were found in the Glacial Sand and Gravel and Boulder Clay respectively (see Figure 6).

Further high-level, often 'very clayey' sandy gravels and silts resting on Boulder Clay have been mapped as Head, where they are predominantly composed of silt grade material, and Head **Gravel** where sand and gravel grades predominate. The occurrences extend from south of Widford south-west to Olives Farm [405 126].

The mineral which ranges in thickness from 3.0 m in IMAU borehole SW 12 [4077 1450] to a total of 14.6 m in SW 14 [4037 1285], has a mean of 6.8 m. Overburden, which ranges from 15.0 m thick in borehole NW 10 [4183 1543] to 1.2 m in borehole SW 21 [4178 1165] has a mean of 6.9 m.

The estimated volume of mineral present in this block is 82 million m³ ±35 per cent at the 95 per cent probability level. Its mean grading is fines 14 per cent, sand 51 per cent and gravel 35 per cent, placing it in the 'clayey' sandy gravel category.

Block C

In block C, which occupies the central part of the resource sheet, all but 5 per cent of the mineral lies beneath a cover of Boulder Clay. Small patches of Glacial Sand and Gravel outcrop in the Fiddlers (Golden) Brook and in the vicinity of Pole Hole [454 127]. Much of the mineral previously exposed in the area of Pole Hole has been extracted and the three pits, covering 0.35 km³, have been restored to agriculture.

To the north, and on both east and west sides of this block, the Boulder Clay overburden thickens and appears to cut down into or through the underlying mineral, as shown by IMAU boreholes NE 17 [4546 1527], NE 23 [4627 1552] and SW 32 [4411 1496]. In borehole SW 30 [4385 1240] there is a thin development of **boulder clay** within the mineral.

Head deposits in Fiddlers (Golden) Brook contain some reworked glacial gravels but they are commonly thin and impersistent and have been assessed as not potentially workable.

The Boulder Clay overburden ranges in recorded thickness from a maximum of 18.0 m in borehole NE 21 [4614 1694] to 1.1 m in SW 37 [4435 1178]: it has a mean of 9.5 m. The mineral, which is entirely glacial in origin, ranges from 3.3 m in SW 37 [4435 1178] to 8.5 m+ in SW 28 [4793 1594] and has a mean of 5.6 m.

The estimated volume of mineral in the block is 65 million m³ ±20 per cent, at the 95 per cent probability level: it has a mean grading of fines 8 per cent, sand 49 per cent and gravel 43 per cent (including 1 per cent of cobble grade) giving an overall mineral classification of sandy gravel.

Block D

This block is divided by the upper reaches of the River Stort whose alluvial, terrace and buried channel deposits have been assessed in block E. In the larger western portion the typical succession of Boulder Clay resting on Glacial Sand and Gravel is maintained. The Boulder Clay contains subordinate beds of sand and gravel: for example in IMAU borehole SE 39 [4756 1446] a thick (8.0 m) 'clayey' pebbly sand lens was encountered high in the Boulder Clay, and similar but smaller lenses are exposed around Thorley Wood [484 185].

The Boulder Clay thickens and cuts down into or replaces the Glacial Sand and Gravel towards the western margin of this block as demonstrated by IMAU boreholes NE 27 [4769 1625], NE 28 [4793 1594] and NE 24 [4768 1961]; in borehole NE 32 [4848 1760] no sand and gravel was present and Boulder Clay rested directly on the Upper Chalk.

East of the River Stort, Head **Gravel** overlies Boulder Clay and Glacial Sand and Gravel in the vicinity of Little Hallingbury [498 180]. The Glacial Sand and Gravel becomes thin or absent on the eastern margin of the sheet and to the south-east, where London Clay is exposed in the valley to the south of Churchgate Street [485 115].

Glacial Sand and Gravel is exposed in a pit at Gaston Green [495 164] where large-scale cross-bedded gravels and sandy gravels can be seen resting on a thin London Clay and mottled clays of the Woolwich and Reading Beds.

Overburden, which comprises mainly Boulder Clay with some Head, ranges in thickness from 0.8 m in borehole NE 38 [4943 1624] to 16.3 m in borehole NE 18 [4667 1953] and has a mean of 6.7 m. Mineral comprising Glacial Sand and Gravel and Head **Gravel**

ranges in recorded thickness from 3.6 m in borehole NE 31 [4842 1884] to 9.8 m in borehole NE 25 [4758 1828] and has a mean of 6.5 m.

The estimate of the volume of mineral is 92 million m^3 ±19 per cent at the 95 per cent probability level. The mineral has a mean grading of fines 11 per cent, sand 54 per cent and gravel 35 per cent, which gives a mineral classification of 'clayey' sandy gravel.

Block E

This block is divided into three mineral-bearing areas associated with the valley of the River Ash in the north-west and the valley of the River Stort in the east and south-west. Both fluvial and glacial deposits have been assessed.

In general the Alluvium and underlying terrace gravels are flanked by small laterally impersistent terraces and Head. In the Ash valley at Hadham Mill the Alluvium overlies terrace gravels and a thick sequence of Glacial Sand and Gravel (see Figure 5). From Bishops Stortford to Churchgate Street in the Stort valley the Alluvium and Head overlie a thick complicated sequence of Boulder Clay, Glacial Sand and Gravel and Lacustrine Deposits which fill a buried tunnel valley.

The overburden ranges in thickness from 0.5 m in borehole NW 14 [4231 1679] to 2.9 m in borehole NW 5 [4029 1573] and has a mean of 1.6 m. The mineral ranges in thickness from 0.7 m in borehole NW 5* to a total of 15.1 m in SE 40 [4790 1302], and has a mean of 4.2 m based on data from IMAU boreholes, commercial records and Hydrogeological Department records.

The block contains an estimated 35 million m^3 of mineral ±47 per cent at the 95 per cent probability level; the mineral has a mean grading of fines 7 per cent, sand 43 per cent and gravel 50 per cent which classifies the mineral as gravel.

---

*Data from this borehole have been used in the calculation of the volume of mineral in this block.

APPENDIX A:

FIELD AND LABORATORY PROCEDURES

Trial and error during initial studies of the complex and variable glacial deposits of East Anglia and Essex showed that an absolute minimum of five sample points evenly distributed across the sand and gravel are needed to provide a worthwhile statistical assessment, but that, where possible, there should be not less than ten. Sample points are any points for which adequate information exists about the nature and thickness of the deposit and may include boreholes other than those drilled during the survey and exposures. In particular, the cooperation of sand and gravel operators ensures that boreholes are not drilled where reliable information is already available; although this may be used in the calculations, it is held confidentially by the Institute and cannot be disclosed.

The mineral shown on each 1:25 000 sheet is divided into resource blocks. The arbitrary size selected, 10 km², is a compromise to meet the aims of the survey by providing sufficient sample points in each block. As far as possible the block boundaries are determined by geological boundaries so that, for example, glacial and river terrace gravels are separated. Otherwise division is by arbitrary lines, which may bear no relationship to the geology. The blocks are drawn provisionally before drilling begins.

A reconnaissance of the ground is carried out to record any exposures and inquiries are made to ascertain what borehole information is available. Borehole sites are then selected to provide an even pattern of sample points at a density of approximately one per square kilometre. However, because broad trends are independently overlain by smaller scale characteristically random variations, it is unnecessary to adhere to a square grid pattern. Thus such factors as ease of access and the need to minimise disturbance to land and the public are taken into account in siting the holes; at the same time it is necessary to guard against the possibility that ease of access (that is, the positions of roads and farms) may reflect particular geological conditions, which may bias the drilling results.

The drilling machine employed should be capable of providing a continuous sample representative of all unconsolidated deposits, so that the in-situ grading can be determined, if necessary, to a depth of 30 m (100 ft) at a diameter of about 200 mm (8 in), beneath different types of overburden. It should be reliable, quiet, mobile and relatively small (so that it can be moved to sites of difficult access). Shell and auger rigs have proved to be almost ideal.

The rigs are modified to enable deposits above the water table to be drilled 'dry', instead of with water added to facilitate the drilling, to minimise the amount of material drawn in from outside the limits of the hole. The samples thus obtained are representative of the in-situ grading, and satisfy one of the most important aims of the survey. Below the water table the rigs are used conventionally, although this may result in the loss of some of the fines fraction and the pumping action of the bailer tends to draw unwanted material into the hole from the sides or the bottom.

A continuous series of bulk samples is taken throughout the sand and gravel. Ideally samples are composed exclusively of the whole of the material encountered in the borehole between stated depths. However, care is taken to discard, as far as possible, material which has caved or has been pumped from the bottom of the hole. A new sample is commenced whenever there is an appreciable lithological change within the sand and gravel, or at every 1 m (3.3 ft) depth. The samples, each weighing between 25 and 45 kg (55 and 100 lb), are despatched in heavy duty polythene bags to a laboratory for grading. The grading procedure is based on British Standard 1377(1967). Random checks on the accuracy of the grading are made in the Institute's laboratories.

All data, including mean grading analysis figures calculated for the total thickness of the mineral, are entered on standard record sheets, abbreviated copies of which are reproduced in Appendix F.

Detailed records may be consulted at the appropriate offices of the Institute, upon application to the Head, Industrial Minerals Assessment Unit.

APPENDIX B

STATISTICAL PROCEDURE

Statistical assessment

1 A statistical assessment is made of an area of mineral greater than 2 km², if there is a minimum of five evenly spaced boreholes in the resource block (for smaller areas see paragraph 12 below).

2 The simple methods used in the calculations are consistent with the amount of data provided by the survey. Conventional symmetrical confidence limits are calculated for the 95 per cent probability level, that is, there is a 5 per cent or one in twenty chance of a result falling outside the stated limits.

3 The volume estimate (V) for the mineral in a given block is the product of the two variables, the sampled areas (A) and the mean thickness ($\bar{l}_m$) calculated from the individual thicknesses at the sample points. The standard deviations for these variables are related such that

$$S_V = \sqrt{(S_A^2 + S_{\bar{l}_m}^2)} \quad . \qquad [1]$$

4 The above relationship may be transposed such that

$$S_V = S_{\bar{l}_m} \sqrt{(1 + S_A^2/S_{\bar{l}_m}^2)} \quad . \qquad [2]$$

From this it can be seen that as $S_A^2/S_{\bar{l}_m}^2$ tends to 0, S_V tends to $S_{\bar{l}_m}$.

If, therefore, the standard deviation for area is small with respect to that for mean thickness, the standard deviation for volume approximates to that for mean thickness.

5 Given that the number of approximately evenly spaced sample points in the sampled area is n with mineral thickness measurements $l_{m_1}, l_{m_2}, \ldots l_{m_n}$, then the best estimate of mean thickness, $\bar{l}_m$, is given by

$$\Sigma(l_{m_1} + l_{m_2} \ldots l_{m_n})/n.$$

For groups of closely spaced boreholes a discretionary weighting factor may be applied to avoid bias (see note on weighting below). The standard deviation for mean thickness $S_{\bar{l}}$, expressed as a proportion of the mean thickness, is given by

$$S_{\bar{l}} = (1/\bar{l}_m)\sqrt{[\Sigma(l_m - \bar{l}_m)^2/(n-1)]}$$

where l_m is any value in the series l_{m_1} to l_{m_n}.

6 The sampled area in each resource block is coloured pink on the map. Wherever possible, calculations relate to the mineral within mapped geological boundaries (which may not necessarily correspond to the limits of deposit). Where the area is not defined by a mapped boundary, that is, where the boundary is inferred, a distinctive symbol is used. Experience suggests that the errors in determining area are small relative to those in thickness. The relationship $S_A/S_{\bar{l}_m} \leqslant \frac{1}{3}$ is assumed in all cases. It follows from equation [2] that

$$S_{\bar{l}_m} \leqslant S_V \leqslant 1.05\, S_{\bar{l}_m}. \qquad [3]$$

7 The limits on the estimate of mean thickness of mineral, $L_{\bar{l}_m}$, may be expressed in absolute units

$\pm(t/\sqrt{n}) \times S_{\bar{l}_m}$ or as a percentage

$\pm(t/\sqrt{n}) \times S_{\bar{l}_m} \times (100/\bar{l}_m)$ per cent, where t is Student's t at the 95 per cent probability level for $(n-1)$ degrees of freedom, evaluated by reference to statistical tables. (In applying Student's t it is assumed that the measurements are distributed normally.)

8 Values of t at the 95 per cent probability level for values of n up to 20 are as follows:

n	t	n	t
1	infinity	11	2.228
2	12.706	12	2.201
3	4.303	13	2.179
4	3.182	14	2.160
5	2.776	15	2.145
6	2.571	16	2.131
7	2.447	17	2.120
8	2.365	18	2.110
9	2.306	19	2.101
10	2.262	20	2.093

(from Table 12, Biometrika Tables for Statisticians, Volume 1, Second Edition, Cambridge University Press, 1962). When n is greater than 20, 1.96 is used (the value of t when n is infinity).

9 In calculating confidence limits for volume, L_V, the following inequality corresponding to equation [3] is applied: $L_{\bar{l}_m} \leq L_V \leq 1.05\, L_{\bar{l}_m}$

10 In summary, for values of n between 5 and 20, L_V is calculated as

$[(1.05 \times t)/\bar{l}_m] \times [\sqrt{\Sigma(l_m - \bar{l}_m)^2/n(n-1)}] \times 100$

per cent, and when n is greater than 20, as

$[(1.05 \times 1.96)/\bar{l}_m] \times [\sqrt{\Sigma(l_m - \bar{l}_m)^2/n(n-1)}] \times 100$

per cent.

11 The application of this procedure to a fictitious area is illustrated in Figures 9 and 10.

Inferred assessment

12 If the sampled area of mineral in a resource block is between 0.25 km^2 and 2 km^2 an assessment is inferred, based on geological and topographical information usually supported by the data from one or two boreholes. The volume of mineral is calculated as the product of the area, measured from field data, and the estimated thickness. Confidence limits are not calculated.

13 In some cases a resource block may include an area left uncoloured on the map, within which mineral (as defined) is interpreted to be generally absent. If there is reason to believe that some mineral may be present, an inferred assessment may be made.

14 No assessment is attempted for an isolated area of mineral less than 0.25 km^2.

15 *Note on weighting* The thickness of a deposit at any point may be governed solely by the position of the point in relation to a broad trend. However, most sand and gravel deposits also exhibit a random pattern of local, and sometimes considerable, variation in thickness. Thus the distribution of sample points need be only approximately regular and in estimating the mean thickness only simple weighting is necessary. In practice, equal weighting can often be applied to thicknesses at all sample points. If, however, there is a distinctly unequal distribution of points, bias is avoided by dividing the sampled area into broad zones, to each of which a value roughly proportional to its area is assigned. This value is then shared between the data points within the zone as the weighting factor.

APPENDIX C

CLASSIFICATION AND DESCRIPTION OF SAND AND GRAVEL

For the purposes of assessing resources of sand and gravel a classification should take account of economically important characteristics of the deposit, in particular the absolute content of fines and the ratio of sand to gravel.

The terminology commonly used by geologists when describing sedimentary rocks (Wentworth, 1922) is not entirely satisfactory for this purpose. For example, Wentworth proposed that a deposit should be described as a 'gravelly sand' when it contains more sand than gravel and there is at least 10 per cent of gravel, provided that there is less than 10 per cent of material finer than sand (less than $\frac{1}{16}$ mm) and coarser than pebbles (more than 64 mm in diameter). Because deposits containing more than 10 per cent fines are not embraced by this system a modified binary classification based on Willman (1942) has been adopted.

When the fines content exceeds 40 per cent the material is not considered to be potentially workable and falls outside the definition of mineral. Deposits which contain 40 per cent fines or less are classified primarily on the ratio of sand to gravel but qualified in the light of the fines content, as follows: less than 10 per cent fines—no qualification; 10 per cent or more but less than 20 per cent fines—'clayey'; 20 to 40 per cent fines—'very clayey'.

The term 'clay' (as written, with single quote marks) is used to describe all material passing $\frac{1}{16}$ mm. Thus it has no mineralogical significance and includes particles falling within the size range of silt. The normal meaning applies to the term clay where it does not appear in single quotation marks.

The ratio of sand to gravel defines the boundaries between sand, pebbly sand, sandy gravel and gravel (at 19:1, 3:1 and 1:1).

Thus it is possible to classify the mineral into one of twelve descriptive categories (see Figure 11). The procedure is as follows:

1 Classify according to ratio of sand to gravel.
2 Describe fines.

For example, a deposit grading 11 per cent gravel, 70 per cent sand and 19 per cent fines is classified as 'clayey' pebbly sand. This short description is included in the borehole log (see Note 11, Appendix D).

Many differing proposals exist for the classification of the grain size of sediments (Atterberg, 1905; Udden, 1914; Wentworth, 1922; Wentworth, 1935; Allen, 1936; Twenhofel, 1937; Lane and others, 1947). As Archer (1970a, b) has emphasised, there is a pressing need for a simple metric scale acceptable to both scientific and engineering interests, for which the class limit sizes correspond closely with certain marked changes in the natural properties of mineral particles. For example, there is an important change in the degree of cohesion between particles at about the $\frac{1}{16}$-mm size, which approximates to the generally accepted boundary between silt and sand. These and other requirements are met by a system based on Udden's geometric scale and a simplified form of Wentworth's terminology (Table 4), which is used in this Report.

The fairly wide intervals in the scale are consistent with the general level of accuracy of the qualitative assessments of the resource blocks. Three sizes of sand are recognised, fine (+$\frac{1}{16}$ –$\frac{1}{4}$ mm), medium (+$\frac{1}{4}$ –1 mm) and coarse (+1 –4 mm). The boundary at 16 mm distinguishes a range of finer gravel (+4 –16 mm), often characterised by abundance of worn tough pebbles of vein-quartz, from larger pebbles often of notably different materials. The boundary at 64 mm distinguishes pebbles from cobbles. The term 'gravel' is used loosely to denote both pebble-sized and cobble-sized material.

The size distribution of borehole samples is determined by sieve analysis, which is presented by the laboratory as logarithmic cumulative curves (see, for example, British

Block Calculation 1:25 000 Block } Fictitious

Area
Block: 11.08 km²
Mineral: 8.32 km²

Mean thickness
Overburden: 2.5 m
Mineral: 6.5 m

Volume
Overburden: 21 million m³
Mineral: 54 million m³

Confidence limits of the estimate of mineral volume at the 95 per cent probability level: ±20 per cent
That is, the volume of mineral (with 95 per cent probability: 54±11 million m³

Thickness estimate measurements in metres
l_o=overburden thickness l_m=mineral thickness

Sample point	Weighting w	Overburden l_o	Overburden wl_o	Mineral l_m	Mineral wl_m	Remarks
SE 14	1	1.5	1.5	9.4	9.4	
SE 18	1	3.3	3.3	5.8	5.8	
SE 20	1	nil	–	6.9	6.9	IMAU
SE 22	1	0.7	0.7	6.4	6.4	boreholes
SE 23	1	6.2	6.2	4.1	4.1	
SE 24	1	4.3	4.3	6.4	6.4	
SE 17	½	1.2	1.6	9.8	7.2	Hydrogeology
123/45	½	2.0		4.6		Unit record
1	¼	2.7		7.3		Close group
2	¼	4.5	2.6	3.2	5.8	of four
3	¼	0.4		6.8		boreholes
4	¼	2.8		5.9		(commercial)
Totals	$\sum w=8$		$\sum wl_o=20.2$		$\sum wl_m=52.0$	
Means		$\bar{l}_o=2.5$		$\bar{l}_m=6.5$		

Calculation of confidence limits

l_m	$(l_m-\bar{l}_m)$	$(l_m-\bar{l}_m)$
9.4	2.9	8.41
5.8	0.7	0.49
6.9	0.4	0.16
6.4	0.1	0.01
4.1	2.4	5.76
6.4	0.1	0.01
7.2	0.7	0.49
5.8	0.7	0.49

$\sum(l_m-\bar{l}_m)^2=15.82$
$n=8$
$t=2.365$

L_V is calculated as

$$1.05\,(t/\bar{l}_m)\sqrt{[\sum(l_m-\bar{l}_m)^2/n(n-1)]}\times 100$$
$$=1.05\times(2.365/6.5)\sqrt{[15.82/(8\times 7)]}\times 100$$
$$=20.3$$
$$\simeq 20 \text{ per cent.}$$

Figure 9 Example of resource block assessment: calculation and results

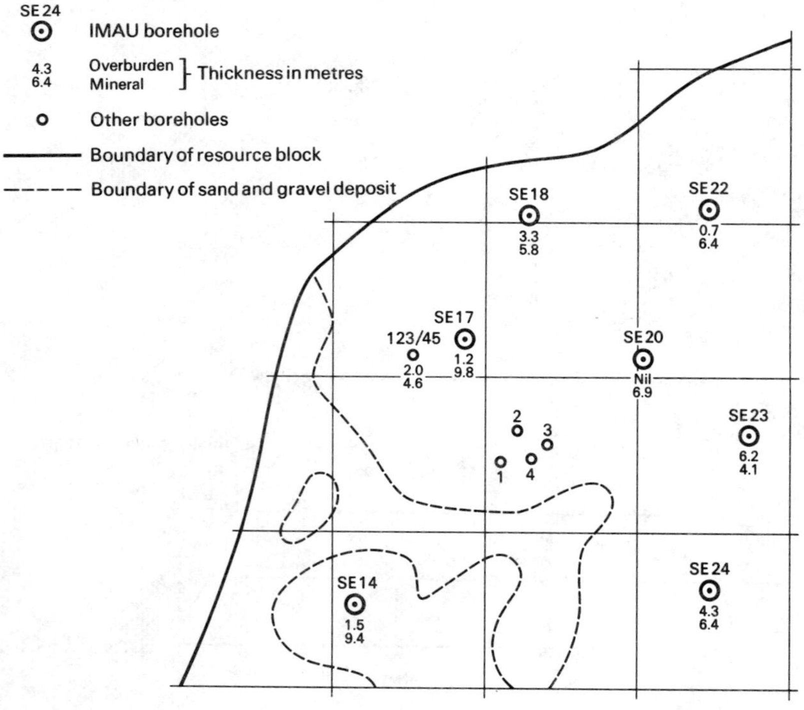

Figure 10 Example of resource block assessment: map of a fictitious block

Standard 1377:1967). In this report the grading is tabulated on the borehole record sheets (Appendix F), the intercepts corresponding with the simple geometric scale $\frac{1}{16}$ mm, $\frac{1}{4}$ mm, 1 mm, 4 mm, 16 mm and so on as required. Original sample grading curves are available for reference at the appropriate office of the Institute.

Each bulk sample is described, subjectively, by a geologist at the borehole site. Being based on visual examination, the description of the grading is inexact, the accuracy depending on the experience of the observer. The descriptions recorded are modified, as necessary, when the laboratory results become available.

The relative proportions of the rock types present in the gravel fraction are indicated by the use of the words 'and' or 'with'. For example, 'flint and quartz' indicates very approximate equal proportions with neither constituent accounting for less than about 25 per cent of the whole; 'flint with quartz' indicates that flint is dominant and quartz, the principal accessory rock type, comprises 5 to 25 per cent of the whole. Where the accessory material accounts for less than 5 per cent of the whole, but is still readily apparent, the phrase 'with some' has been used. Rare constituents are referred to as 'trace'.

The terms used in the field to describe the degree of rounding of particles, which is concerned with the sharpness of the edges and corners of a clastic fragment and not the shape (after Pettijohn, 1957), are as follows.

Angular: showing little or no evidence of wear; sharp edges and corners.

Subangular: showing definite effects of wear. Fragments still have their original form but edges and corners begin to be rounded off.

Subrounded: showing considerable wear. The edges and corners are rounded off to smooth curves. Original grain shape is still distinct.

Rounded: original faces almost completely destroyed, but some comparatively flat surfaces may still remain. All original edges and corners have been smoothed off to rather broad curves. Original shape is still apparent.

Well-rounded: no original faces, edges or corners left. The entire surface consists of broad curves; flat areas are absent. The original shape is suggested by the present form of the grain.

Table 4 Classification of gravel, sand and fines

Size limits	Grain size description	Qualification	Primary classification
	Cobble		
64 mm –		Coarse	Gravel
16 mm –	Pebble		
		Fine	
4 mm –			
		Coarse	
1 mm –			
	Sand	Medium	Sand
$\frac{1}{4}$ mm –			
		Fine	
$\frac{1}{16}$ mm –			
	Fines (silt and clay)		Fines

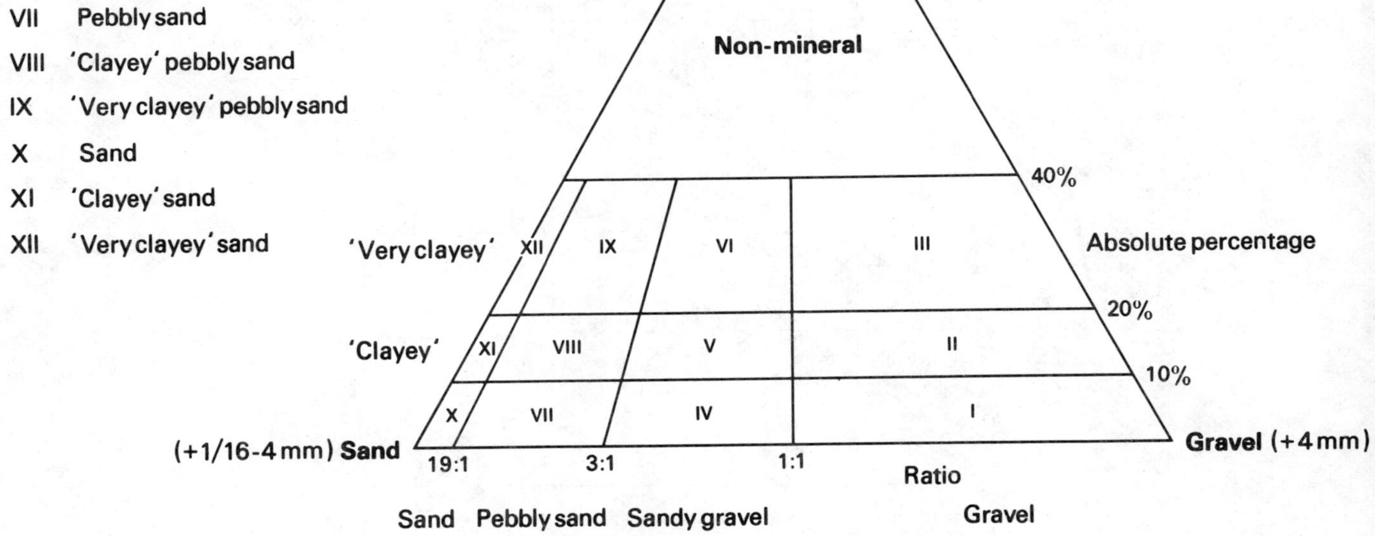

Figure 11 Diagram to show the descriptive categories used in the classification of sand and gravel

APPENDIX D

EXPLANATION OF THE BOREHOLE RECORDS

ANNOTATED EXAMPLE

TL 41 NE 29[1] **4710 1548[2]** **Tharbies Lodge, Sawbridgeworth[3]** **Block D**

Surface level + 68.2 m[4]
Water not struck[5]
Shell 152 mm diameter[6]
October 1975

Overburden[7] 11.4 m
Mineral 4.2 m
Waste 0.3 m
Bedrock 6.5 m+[9]

Log

Geological classification[10]	*Lithology*[11]	*Thickness* m	*Depth*[8] m
	Soil	0.3	0.3
Boulder Clay	Clay, very silty, chalky, pebbly, yellow brown, soft	3.6	3.9
	Clay, chalky, silty, blue grey becoming dark grey, with some flint and quartz pebbles, firm becoming stiff	7.0	10.9
	Clay, silty, sandy, very pebbly, dark brown, firm	0.5	11.4
Glacial Sand and Gravel	'Clayey' gravel Sand: medium and coarse with fine, predominantly quartz, with some coarse angular flints, yellow brown Gravel: fine and coarse, subrounded to angular flint, with some rounded quartz and quartzite, and a trace of sandstone	4.2	15.6
	Clay, sandy, pebbly, yellow brown, soft	0.3	15.9
London Clay (Basement Bed)	Sand, very silty, clayey, fine with some medium, laminations in parts, some shell debris, poorly developed horizon of rounded brown and black flint pebbles at base, drab olive green	5.9	21.8
Woolwich and Reading Beds	Clay, mottled blue, red and brown, stiff, becoming hard, waxy	0.6+	22.4

Grading[13]

Mean for deposit[14] *percentages*			Depth below surface (m)	*percentages*					
Fines	Sand	Gravel		Fines	Sand			Gravel	
				$-\frac{1}{16}$	$+\frac{1}{16}-\frac{1}{4}$	$+\frac{1}{4}-1$	+1–4	+4–16	+16
11	40	49	[12]11.4–12.4	10	7	19	12	29	23
			12.4–13.4	9	7	20	12	22	30
			13.4–14.4	11	9	19	12	25	24
			14.4–15.6	12	10	19	14	28	17
			Mean	11	8	19	13	26	23

Composition[15]

Depth below surface (m)	Percentages by weight in + 4–16 mm fraction					
	Chalk	Flint	Quartz and Quartzite	Fossil debris	Sandstone	Others
11.4–12.4	–	79	14	–	4	3
12.4–13.4	–	75	14	–	7	4
13.4–14.4	–	71	18	–	7	4
14.4–15.6	–	68	21	–	7	4
Mean	–	73	17	–	6	4

The numbered paragraphs below correspond with the annotations given on the specimen record above.

1 Borehole Registration Number
Each Industrial Minerals Assessment Unit (IMAU) borehole is identified by a Registration Number. This consists of two statements.

a The number of the 1:25 000 sheet on which the borehole lies, that is TL 41

b The quarter of the 1:25 000 on which the borehole lies and its number in a series for that quarter, for example NE 29. Thus the full Registration Number is TL 41 NE 29. Usually this is abbreviated to NE 29 in the text.

2 National Grid reference
All National Grid references in this publication lie within the 100-km square TL. Grid references are given to eight figures, accurate to within 10 m for borehole location and pit exposures. (In the text six-figure grid references are used for more approximated location, for example farms, woods etc.)

3 Location
The position of the borehole is generally referred to the nearest named locality on the 1:25 000 base map. The parish and the block in which it lies is also given.

4 Surface level
The surface level at the borehole site is given in metres above Ordnance Datum.

5 Groundwater conditions
If groundwater was present the level at which it was encountered is normally given (in metres above Ordnance Datum). If no water was encountered the words 'water not struck' are used.

6 Type of drill and date of drilling
The type of machine, the external diameter of the casing used, and the month and year in which drilling took place are stated.

7 Overburden, mineral, waste and bedrock
Mineral is sand and gravel which, as part of a deposit, falls within the arbitrary definition of potentially workable material (see p. 1). Bedrock is the 'formation', 'country rock' or 'rock head' below which potentially workable sand and gravel will not be found. Waste is any material other than bedrock or mineral. Where waste occurs between the surface and mineral it is classified as overburden.

8 Thickness and depth
All measurements were made in metres. Conversions to imperial measure have been omitted from the logs but a table of conversions metres to feet (rounded to the nearest 0.5 ft) is given in Appendix H.

9 The plus (+) sign indicates that the base of the deposit was not reached during drilling.

10 Geological classification
The geological classification is given wherever possible.

11 Lithological description
When sand and gravel is recorded a general description based on the mean grading characteristics (for details see Appendix C) is followed by more detailed particulars. The description of other rocks is based on visual examination in the field.

12 Sampling
A continuous series of bulk samples is taken throughout the thickness of sand and gravel. A new sample is commenced whenever there is an appreciable lithological change within the sand and gravel or at every 1 m of depth. Samples obtained by the bailing technique (that is, from deposits below the water table) are indicated by an asterisk.

13 Grading results
The limits are as follows: gravel, +4 mm; sand, –4 mm + $\frac{1}{16}$ mm; fines, – $\frac{1}{16}$ mm. A fuller classification is given in Table 4. No attempt has been made to estimate the grading where, exceptionally, results are not available.

14 Mean grading
The grading of the full thickness of the mineral identified in the log is the mean of the individual sample gradings weighted by the thicknesses represented, if these vary.

Fully representative sampling of sand and gravel is difficult to achieve particularly where groundwater levels are high. Comparison between boreholes and adjacent exposures suggests that in borehole samples the proportion of sand may be higher and the proportion of fines and coarse gravel (+16 mm) may be lower.

15 Composition
The composition of the fine gravel fraction (+4 –16 mm) is given. Material of metamorphic, igneous and sedimentary origin, which individually occurs rarely (for example, schist, granite, basalt, siltstone, limestone and mudstone) is included in the column headed 'Others'. The mean composition is quoted for the full mineral thickness.

APPENDIX E:

LIST OF BOREHOLES USED IN THE ASSESSMENT OF RESOURCES

Borehole number*	Grid reference†
IMAU BOREHOLES	
NW 1	4036 1954
2	4050 1862
3	4014 1741
4	4068 1683
5	4029 1573
6	4059 1504
7	4147 1893
8	4143 1814
9	4186 1627
10	4183 1543
11	4232 1989
12	4239 1856
13	4213 1783
14	4231 1679
15	4281 1531
16	4312 1983
17	4338 1875
18	4370 1818
19	4319 1751
20	4378 1675
21	4381 1604
22	4436 1978
23	4480 1867
24	4440 1775
25	4465 1670

Borehole number*	Grid reference†
NW 26	4498 1579
27	4408 1557
NE 13	4570 1976
14	4532 1891
15	4589 1850
16	4539 1795
17	4546 1527
18	4667 1953
19	4678 1835
20	4684 1747
21	4614 1694
22	4663 1660
23	4627 1552
24	4768 1961
25	4758 1828
26	4764 1729
27	4769 1625
28	4793 1594
29	4710 1548
30	4829 1940
31	4842 1884
32	4848 1760
33	4856 1676
34	4967 1915
35	4920 1867

Borehole number*	Grid reference†
NE 36	4977 1813
37	4939 1732
38	4943 1624
39	4917 1505
SW 12	4077 1450
13	4076 1372
14	4037 1285
15	4067 1226
16	4048 1151
17	4072 1065
18	4182 1372
19	4172 1289
20	4172 1289
21	4178 1165
22	4135 1062
23	4262 1449
24	4259 1343
25	4272 1236
26	4245 1118
27	4239 1034
28	4388 1437
29	4328 1308
30	4385 1240
31	4371 1143
32	4411 1496

Borehole number*	Grid reference†
SW 33	4464 1404
34	4421 1353
35	4417 1349
36	4472 1238
37	4435 1178
SE 34	4543 1464
35	4553 1361
36	4570 1227
37	4667 1431
38	4660 1338
39	4756 1446
40	4790 1302
41	4770 1061
42	4847 1404
43	4875 1319
44	4862 1220
45	4876 1109
46	4837 1030
47	4961 1455
48	4956 1452
49	4963 1351
50	4922 1285
51	4985 1089

* By sheet quadrant
† All fall within TL 41

HYDROGEOLOGICAL DEPARTMENT RECORDS‡ (all fall within TL 41)
NW: 28, 29, 30, 34, 35, 41, 43, 44, 50, 52, 59.
NE: 47, 48, 50, 51, 52, 53, 57, 58, 59, 60, 62, 64.
SW: 39, 42, 43, 44, 46, 47, 51, 56.
SE: 70, 71, 78, 80.

RECORDS FROM OTHER SOURCES

Information from 85 sample points (boreholes and trial pits) was also used in the assessment. This information was obtained from Local and County Councils, the minerals industry, private individuals and from the Institute's borehole register.

‡ The Hydrogeological Department records have now been incorporated in the Institute's borehole register. The new registered numbers are quoted above.

APPENDIX F:

INDUSTRIAL MINERALS ASSESSMENT UNIT BOREHOLE RECORDS

TL 41 NW 1 4036 1954 Bartram Wood, Standon **Block A**

Surface level +85.5 m
Water not struck
Shell (modified) 152 mm diameter
February 1976

Waste 18.7 m+

Log

Geological classification	*Lithology*	*Thickness* m	*Depth* m
	Soil	0.4	0.4
Boulder Clay	Clay, silty, chalky, sandy, brown, with some angular flints, firm	2.6	3.0
	Clay, silty, sandy, grey, stiff	1.0	4.0
	Clay, silty, chalky, dark grey becoming dark olive grey, with some angular flints and rounded quartz pebbles, stiff	11.2	15.2
	Clay, silty, chalky, yellow grey becoming dark grey, chalky sand bands in upper 0.5 m, some flint and quartz with a trace of Jurassic material, stiff	3.5+	18.7

TL 41 NW 2 4050 1862 The Nursery, Much Hadham **Block A**

Surface level +87.1 m
Water struck at +70.8 m
Shell and Auger 152 mm diameter
June 1972

Overburden 11.3 m
Mineral 7.0 m
Bedrock 0.9 m+

Log

Geological classification	*Lithology*	*Thickness* m	*Depth* m
	Soil	0.1	0.1
Boulder Clay	Clay, brown, with angular flints and chalk fragments, friable	0.4	0.5
	Clay, silty, chalky, grey, with some flint, soft	2.7	3.2
	Clay, chalky, grey, soft	3.9	7.1
	Clay, chalky, brown, soft	0.2	7.3
	'Very clayey' sand, medium with some fine, brown, soft	0.4	7.7
	Clay, chalky, grey, soft	3.2	10.9
	Clay, chalky, pebbly, brown, soft	0.4	11.3
Glacial Sand and Gravel	'Very clayey' sandy gravel, pebbly sand in upper 3.0 m becomes progressively more iron-cemented with depth Sand: medium with some coarse and fine, brown Gravel: fine with coarse, subangular to angular flint, with some subrounded to well rounded quartz, quartzite and sandstone	7.0	18.3
	Clay, brown, with angular flints and small subangular quartz pebbles, soft	0.8	19.1
Upper Chalk	Chalk, white, soft	0.1+	19.2

Grading

Mean for deposit *percentages*			Depth below surface (m)	*percentages*					
Fines	Sand	Gravel		Fines	Sand			Gravel	
				$-\frac{1}{16}$	$+\frac{1}{16}-\frac{1}{4}$	$+\frac{1}{4}-1$	+1–4	+4–16	+16
28	71	1	7.3–7.7	28	12	57	2	1	–
22	55	23	11.3–12.3	No sample					
			12.3–13.3	17	4	67	3	7	2
			13.3–14.3	32	4	53	4	6	1
			14.3–15.3	34	2	32	2	8	22
			15.3–16.3	24	5	39	5	10	17
			16.3–17.3*	12	6	22	20	36	4
			17.3–18.3*	15	12	29	19	21	4
			Mean	22	6	40	9	15	8

TL 41 NW 3 4014 1741 Nimney Brook, Ware Rural **Block A**

Surface level +65.9 m
Water not struck
Shell 152 mm diameter
October 1975

Overburden 2.7 m
Mineral 3.3 m
Bedrock 0.5 m+

Log

Geological classification	*Lithology*	*Thickness* m	*Depth* m
	Soil	0.2	0.2
Boulder Clay	Clay, sandy, chalky, brown, with some flint, firm	2.5	2.7
Glacial Sand and Gravel	'Very clayey' sandy gravel, becomes more gravelly with depth Sand: medium with fine and some coarse, predominantly quartz, brown Gravel: Coarse and fine, subrounded to rounded flints, with a trace of angular flint, and some rounded quartz and quartzite, with a trace of sandstone	3.3	6.0
Upper Chalk	Chalk, pugged up, white streaked brown, with occasional black and blue flints	0.5+	6.5

Grading

Mean for deposit *percentages*			Depth below surface (m)	*percentages*					
Fines	Sand	Gravel		Fines	Sand			Gravel	
				$-\frac{1}{16}$	$+\frac{1}{16}-\frac{1}{4}$	$+\frac{1}{4}-1$	+1–4	+4–16	+16
18	46	36	2.7–3.7	31	10	27	5	11	16
			3.7–4.7	9	24	33	7	11	16
			4.7–5.7	16	10	13	8	25	28
			5.7–6.0	No sample					
			Mean	18	15	24	7	16	20

Composition

Depth below surface (m)	Percentages by weight in + 4–16 mm fraction					
	Chalk	Flint	Quartz and Quartzite	Fossil debris	Sandstone	Others
2.7–3.7	–	77	14	–	3	6
3.7–4.7	–	63	21	–	10	6
4.7–5.7	–	92	6	–	1	1
Mean	–	77	14	–	5	4

TL 41 NW 4 **4068 1683** **Sheepcote Plantation, Ware Rural** **Block A**

Surface level +73.1 m
Water not struck
Shell 152 mm diameter
September–October 1975

Overburden 9.9 m
Mineral 4.1 m
Bedrock 2.0 m+

Log

Geological classification	*Lithology*	*Thickness* m	*Depth* m
	Soil	0.2	0.2
Boulder Clay	Clay, chalky, silty, pale blue, soft	1.1	1.3
	Clay, chalky, silty, yellow brown becoming grey blue, with some flint and quartz pebbles, firm becoming stiff	8.1	9.4
	Clay, pebbly, chalky, sandy, silty, brown, firm	0.5	9.9
Glacial Sand and Gravel	'Clayey' sandy gravel, becomes 'very clayey' and less gravelly with depth Sand: medium with coarse and fine, predominantly quartz with angular coarse flint Gravel: fine and coarse with a trace of cobble, angular and rounded black and brown flints, with rounded quartz, quartzite and sandstone, and a trace of rounded chalk	4.1	14.0
Woolwich and Reading Beds	Clay, very silty, sandy, faintly laminated, mottled red, yellow, green and brown, with a trace of rounded black flints and very rare rounded cream quartz at base, firm	1.3	15.3
Thanet Beds (Bullhead Bed)	Sand, fine, dark olive green, glauconitic with many angular black flints, firm	0.3	15.6
Upper Chalk	Chalk, white, streaked brown, gritty, pugged up, soft	0.4+	16.0

Grading

Mean for deposit percentages			Depth below surface (m)	*percentages*						
Fines	Sand	Gravel		Fines	Sand			Gravel		
				$-\frac{1}{16}$	$+\frac{1}{16}-\frac{1}{4}$	$+\frac{1}{4}-1$	+1–4	+4–16	+16–64	+64
15	39	46	9.9–10.9	5	6	17	13	25	34	–
			10.9–11.9	12	8	14	15	25	22	4
			11.9–12.9	14	11	18	11	24	22	–
			12.9–14.0	27	13	19	9	20	12	–
			Mean	15	10	17	12	23	22	1

Composition

Depth below surface (m)	Percentages by weight in + 4–16 mm fraction					
	Chalk	Flint	Quartz and Quartzite	Fossil debris	Sandstone	Others
9.9–10.9	3	52	25	–	14	6
10.9–11.9	1	69	18	–	11	1
11.9–12.9	4	68	12	–	11	5
12.9–14.0	7	65	16	–	8	4
Mean	4	63	18	–	11	4

TL 41 NW 5 4029 1573 Wareside Lodge, Ware Rural **Block E**

Surface level +40.8 m
Water struck at +37.8 m
Shell 152 mm diameter
October 1975

Overburden 2.9 m
Mineral 0.7 m
Bedrock 0.1 m+

Log

Geological classification	*Lithology*	*Thickness* m	*Depth* m
	Soil	0.1	0.1
Alluvium	Clay, pebbly, mid-brown, soft	0.4	0.5
	Clay, sandy, silty, chalky, pebbly, brown streaked grey becoming blue and carbonaceous	2.4	2.9
	'Very clayey' gravel Sand: coarse angular flint with a trace of medium and fine, grey Gravel: fine and coarse, rounded to well rounded flint with rounded quartz	0.7	3.6
Upper Chalk	Chalk, white streaked brown, pugged up, with fresh angular black flints	0.1+	3.7

TL 41 NW 6 4059 1504 Thistly Wood, Widford **Block B**

Surface level +77.4 m
Water struck at +59.4 m
Shell and Auger 152 mm diameter
June–July 1972

Overburden 13.3 m
Mineral 6.2 m
Bedrock 2.3 m+

Log

Geological classification	*Lithology*	*Thickness* m	*Depth* m
	Soil	0.3	0.3
Boulder Clay	Clay, chalky, grey, soft	12.5	12.8
	Clay, chalky, sandy, pebbly, brown, soft	0.5	13.3
Glacial Sand and Gravel	Gravel: Sand: medium with coarse and a trace of fine, predominantly angular flint with some rounded quartz Gravel: fine and coarse, subangular to angular flint, with some subrounded to well rounded flint quartz and quartzite	6.2	19.5
London Clay	Clay, silty, grey, stiff	2.3	21.8
Woolwich and Reading Beds	Clay, mottled blue and green, stiff, waxy	Just Penetrated	

Grading

Mean for deposit *percentages*			Depth below surface (m)	*percentages*					
Fines	Sand	Gravel		Fines	Sand			Gravel	
				$-\frac{1}{16}$	$+\frac{1}{16}-\frac{1}{4}$	$+\frac{1}{4}-1$	+1–4	+4–16	+16
9	43	48	14.2–15.2	5	2	16	13	45	19
			15.2–16.2	18	3	49	8	17	5
			16.2–17.2	No sample					
			17.2–18.0	10	3	30	13	21	23
			18.0–19.0*	2	1	16	16	28	37
			19.0–19.5*	15	3	28	18	20	16
			Mean	9	2	28	13	28	20

TL 41 NW 7 4147 1893 Brands Farm, Much Hadham **Block A**

Surface level +93.4 m
Water not struck
Shell (modified) 152 mm diameter
February 1976

Waste 18.8 m+

Log

Geological classification	*Lithology*	*Thickness* m	*Depth* m
	Soil	0.3	0.3
Boulder Clay	Clay, extremely chalky, silty, yellow grey, with flint pebbles, firm	5.3	5.6
	Clay, silty, chalky, dark grey, stiff	2.8	8.4
	Clay, silty, chalky, dark grey to olive grey, with flint, sandstone, mudstone, quartz and quartzite pebbles, dark grey shelly limestone boulder between 12.4 and 12.6 m	9.8	18.2
	Clay, silty, chalky, pebbly, grey brown, stiff	0.4	18.6
Glacial Sand and Gravel	'Very clayey' sandy gravel Sand: medium to coarse with some fine, predominantly angular flint, orange brown Gravel: coarse and cobble sized well rounded quartz, quartzite and flint, with some angular flint	0.2+	18.8

TL 41 NW 8 **4143 1814** **Moat Farm, Much Hadham** **Block A**

Surface level +90.1 m
Water struck at +81.6 m
Shell (modified) 152 mm diameter
February 1976

Waste 25.0 m+

Log

Geological classification	*Lithology*	*Thickness* m	*Depth* m
	Soil	0.1	0.1
Boulder Clay	Clay, chalky, silty, brown, soft	1.6	1.7
	Clay, chalky, pebbly, grey brown, soft	0.2	1.9
	Sand, very clayey, pebbly, brown, firm	0.1	2.0
	Clay, chalky, silty, brown to grey, some flint, soft	3.1	5.1
	Clay, silty, chalky, blue grey, some flint and black shale, stiff	3.4	8.5
	Clay, chalky, sandy, brown, soft	0.1	8.6
	Clay, silty, chalky, with some flint pebbles, blue grey, stiff	2.9	11.5
	Clay, chalky, silty, dark blue grey with dark brown tinge at depth, stiff	6.0	17.5
	Clay, silty, very sandy, ochre brown, soft	0.6	18.1
	Clay, extremely pebbly and sandy, brown, firm	0.6	18.7
	Clay, silty, sandy, white flints, orange,firm	0.6	19.3
Glacial Sand and Gravel	Sandy gravel, 'very clayey' to 'clayey' in upper 2.0 m Sand: medium with some coarse and fine, predominantly rounded quartz,with angular coarse flint, orange brown Gravel: fine with coarse, subrounded to angular flint,with some quartz, quartzite and sandstone	5.7+	25.0

Grading

Mean for deposit *percentages*			Depth below surface (m)	*percentages*					
Fines	Sand	Gravel		Fines	Sand			Gravel	
				$-\frac{1}{16}$	$+\frac{1}{16}-\frac{1}{4}$	$+\frac{1}{4}-1$	+1–4	+4–16	+16
9	52	39	19.3–20.0	22	17	57	2	2	–
			20.0–21.0	10	6	25	11	27	21
			21.0–22.0	9	5	28	11	28	19
			22.0–23.0	7	12	31	7	24	19
			23.0–24.0	6	6	40	7	24	17
			24.0–25.0	6	7	40	9	23	15
			Mean	9	8	36	8	23	16

Composition

Depth below surface (m)	Percentages by weight in + 4–16 mm fraction				
	Chalk	Flint	Quartz and Quartzite	Fossil debris	Sandstone	Others
19.3–20.0	–	71	18	–	5½	5½
20.0–21.0	–	70	18	–	9	3
21.0–22.0	–	72	23	–	4	1
22.0–23.0	–	75	12	–	8	5
23.0–24.0	–	73	17	–	8	2
24.0–25.0	–	61	22	–	11	6
Mean	–	70	18	–	8	4

TL 41 NW 9 4186 1627 Pegs Lane, Widford **Block A**

Surface level +47.9 m
Water not struck
Shell 152 mm diameter
October 1975

Waste 0.7 m
Bedrock 1.3 m+

Log

Geological classification	*Lithology*	*Thickness* m	*Depth* m
	Soil	0.2	0.2
Head	Clay, silty, pebbly, brown, friable	0.5	0.7
Upper Chalk	Chalk, pugged up, white streaked brown	1.3+	2.0

TL 41 NW 10 **4183 1543** **Abbots Farm, Widford** **Block B**

Surface level +82.2 m
Water not struck
Shell 152 mm diameter
October 1975

Overburden 15.0 m
Mineral 7.2 m
Bedrock 1.8 m+

Log

Geological classification	*Lithology*	*Thickness* m	*Depth* m
	Soil	0.1	0.1
Boulder Clay	Clay, very sandy, very pebbly, brown, with flint and quartz pebbles, firm	1.4	1.5
	Clay, pebbly, chalky, silty, brown, firm	3.5	5.0
	Clay, chalky, blue grey, with some flint and quartz pebbles, stiff	9.7	14.7
	Clay, chalky, pebbly, sandy, brown, firm	0.3	15.0
Glacial Sand and Gravel	'Clayey' sandy gravel, fine to medium sand in topmost 1.0 m, becoming progressively more pebbly in next 1.4 m, becomes more clayey and coarser with depth Sand: medium and fine with some coarse, predominantly rounded quartz, with angular coarse and medium flint, yellow becoming brown Gravel: fine with coarse and a trace of cobble, subangular to rounded flint, with rounded to subrounded quartz, quartzite and sandstone, and a trace of chalk	7.2	22.2
Woolwich and Reading Beds	Clay, sandy becoming very sandy, green mottled brown and rarely red, firm	0.7	22.9
	Sand, clayey, buff, fine and medium, soft	1.1+	24.0

Grading

Mean for deposit *percentages*			Depth below surface (m)	*percentages*						
Fines	Sand	Gravel		Fines	Sand			Gravel		
				$-\frac{1}{16}$	$+\frac{1}{16}-\frac{1}{4}$	$+\frac{1}{4}-1$	+1–4	+4–16	+16–64	+64
14	59	27	15.0–16.0	11	65	23	1	–	–	–
			16.0–17.0	11	27	53	4	2	3	–
			17.0–17.4	19	19	39	5	7	6	5
			17.4–18.4	20	13	22	12	24	9	–
			18.4–19.4	16	13	20	16	25	10	–
			19.4–20.4	18	14	18	15	22	13	–
			20.4–21.2	19	15	20	15	20	11	–
			21.2–22.2	6	10	17	16	25	24	2
			Mean	14	22	26	11	16	10	1

Composition

Depth below surface (m)	Percentages by weight in + 4–16 mm fraction					
	Chalk	Flint	Quartz and Quartzite	Fossil debris	Sandstone	Others
15.0–16.0	Sand no pebbles					
16.0–17.0	9	47	18	–	19	7
17.0–17.4	12	59	17	–	7	5
17.4–18.4	2	71	16	–	10	1
18.4–19.4	–	81	12	–	5	2
19.4–20.4	–	80	14	–	5	1
20.4–21.2	–	79	13	–	6	2
21.2–22.2	–	79	12	–	7	2
Mean	2	72	14	–	9	3

TL 41 NW 11 4232 1989 New Barns, Much Hadham **Block A**

Surface level +80.2 m
Water not struck
Shell (modified) 152 mm diameter
February 1976

Waste 12.5 m
Bedrock 2.0 m+

Log

Geological classification	*Lithology*	*Thickness* m	*Depth* m
	Soil	0.2	0.2
Boulder Clay	Clay, silty, chalky, orange brown mottled grey, soft	2.2	2.4
	Clay, chalky, dark grey, firm	2.1	4.5
	Silt, fine sandy, olive brown, soft	0.6	5.1
	Clay, silty, chalky, olive brown becoming grey and brown, stiff	1.1	6.2
Glacial Sand and Gravel	**a** 'Very clayey' sandy gravel Sand: medium and fine with some coarse, predominantly flint, and quartz, with chalk, white Gravel: fine with some coarse, angular to well rounded flint, with some rounded chalk and a trace of quartz, quartzite sandstone and fossil debris	1.5	7.7
Boulder Clay	Clay, chalky, silty, flinty, pale yellow, firm	1.8	9.5
	Gravel, very silty, clayey, sandy angular flints with rare rounded quartz and chalk	0.2	9.7
	Clays and silts laminated, fine sandy, chalky, olive and grey becoming brown, orange and yellow, firm	1.0	10.7
Glacial Sand and Gravel	**b** 'Clayey' gravel Sand: medium and fine with some coarse, predominantly quartz, with angular coarse flint Gravel: fine and coarse with some cobble, angular to subangular flint, with rounded quartz, quartzite, and sandstone	1.8	12.5
Woolwich and Reading Beds	Silt, fine sandy, clayey, bioturbated olive grey streaked pale grey, soft	0.7	13.2
	Clay, sandy, orange, with angular and well rounded black flints	0.8	14.0
Upper Chalk	Chalk, white, soft	0.5+	14.5

Grading

	Mean for deposit *percentages*			Depth below surface (m)	*percentages*						
	Fines	Sand	Gravel		Fines	Sand			Gravel		
					$-\frac{1}{16}$	$+\frac{1}{16}-\frac{1}{4}$	$+\frac{1}{4}-1$	+1–4	+4–16	+16–64	+64
a	32	45	23	6.2–7.2	33	17	15	9	18	8	–
				7.2–7.7	29	17	27	8	13	6	–
				Mean	32	17	19	9	16	7	–
b	10	40	50	10.7–11.7	12	12	17	8	22	16	13
				11.7–12.5	7	18	16	8	21	30	–
				Mean	10	15	17	8	21	22	7

Composition

	Depth below surface (m)	Percentages by weight in +4–16 mm fraction					
		Chalk	Flint	Quartz and Quartzite	Fossil debris	Sandstone	Others
a	6.2–7.2	40	47	6	1	3	3
	7.2–7.7	36	55	5	–	3	1
	Mean	39	49	6	1	3	2
b	10.7–11.7	–	61	30	–	4	5
	11.7–12.5	–	74	11	–	12	3
	Mean	–	67	22	–	7	4

TL 41 NW 12   4239 1856   Moor Place, Much Hadham   **Block A**

Surface level +68.9 m
Water struck at +64.3 m and +52.4 m
Shell (modified) 152 mm diameter
March 1976

Waste 18.6 m+

Log

Geological classification	*Lithology*	*Thickness* m	*Depth* m
	Soil	0.3	0.3
Boulder Clay	Clay, chalky, silty, flinty, yellow brown, firm	4.3	4.6
	Clay, silty, chalky, firm to stiff, with a chalk sand band at 4.7 m, grey becoming dark grey	11.9	16.5
	'Clayey' sandy gravel, chalky medium and coarse sand and fine chalk gravel, grey	0.2	16.7
	Clay, chalky, silty, grey, stiff	1.9+	18.6

TL 41 NW 13 4213 1783 Wynches, Much Hadham **Block A**

Surface level +78.2 m
Water not struck
Shell (modified) 152 mm diameter
February 1976

Overburden 4.8 m
Mineral 7.0 m
Bedrock 2.0 m+

Log

Geological classification	*Lithology*	*Thickness* m	*Depth* m
	Soil	0.3	0.3
Boulder Clay	Clay, silty, pebbly, sandy, brown becomes yellow brown, soft	1.9	2.2
	'Very clayey' fine sand; chalky, yellow brown, with some flint pebbles, soft	0.6	2.8
	Clay, chalky, silty, brown, soft	0.3	3.1
	Clay, silty, chalky, grey blue, soft becoming stiff	1.4	4.5
	Clay, silty, chalky, dark brown, with flint pebbles, stiff	0.3	4.8
Glacial Sand and Gravel	Gravel, sand between 10.6 and 10.9 m Sand: medium with coarse and fine, predominantly rounded quartz with some angular coarse and medium flint, brown to orange brown Gravel: fine and coarse with a trace of cobble, angular to subrounded flint, with some rounded quartz and quartzite, and a trace of sandstone	7.0	11.8
Woolwich and Reading Beds	Clay, silty, sandy, brown, firm	0.2	12.0
	Clay, sandy and glauconitic in lenses, mottled red brown and grey becomes grey at base, with some calcareous nodules increasing with depth, stiff	1.3	13.3
	Clay, becomes very clayey, very sandy silt, grey becomes dark purple grey, calcareous nodules in clay, with many red, black and brown well rounded flint pebbles, and a trace of white quartz at base	0.7+	14.0

Grading

Mean for deposit *percentages*			Depth below surface (m)	*percentages*						
Fines	Sand	Gravel		Fines	Sand			Gravel		
				$-\frac{1}{16}$	$+\frac{1}{16}-\frac{1}{4}$	$+\frac{1}{4}-1$	+1–4	+4–16	+16–64	+64
6	42	52	4.8–5.8	7	14	19	10	31	19	–
			5.8–6.8	5	8	12	12	23	26	14
			6.8–7.8	5	10	21	10	27	27	–
			7.8–8.8	6	14	21	9	26	24	–
			8.8–9.8	7	8	18	15	27	25	–
			9.8–10.6	5	6	21	15	25	28	–
			10.6–11.8	8	5	34	10	25	18	–
			Mean	6	9	21	12	26	24	2

Composition

Depth below surface (m)	Percentages by weight in + 4–16 mm fraction					
	Chalk	Flint	Quartz and Quartzite	Fossil debris	Sandstone	Others
4.8–5.8	–	63	29	–	5	3
5.8–6.8	–	67	27	–	5	1
6.8–7.8	–	59	34	–	6	1
7.8–8.8	–	68	29	–	1½	1½
8.8–9.8	–	64	29	–	6	1
9.8–10.6	–	67	27	–	4	2
10.6–11.8	–	70	25	–	3	2
Mean	–	66	29	–	4	1

TL 41 NW 14 4231 1679 Hadham Mill, Much Hadham **Block E**

Surface level +51.9 m
Water struck at +44.9 m
Shell (modified) 152 mm diameter
January 1976

Overburden 0.5 m
Mineral 14.5 m
Bedrock 0.5 m+

Log

Geological classification	*Lithology*	*Thickness* m	*Depth* m
	Soil and very pebbly subsoil	0.5	0.5
Terrace 1	**a** Gravel, with some thin clay bands throughout Sand: coarse and medium with some fine, predominantly rounded quartz, with coarse and medium angular flint, yellow brown Gravel: coarse and fine, angular and rounded flint,with some subrounded to rounded quartz,and a trace of sandstone	4.0	4.5
Glacial Sand and Gravel	**b** Gravel, clayey in upper 3.0 m Sand: medium and coarse with some fine, predominantly angular flint, with some rounded quartz and chalk, becomes angular flint with rounded chalk, and some rounded quartz, pale yellow brown Gravel: fine and coarse, angular to rounded flint, with some rounded chalk, quartz and quartzite, with a trace of sandstone and very rare fossil debris	10.5	15.0
Upper Chalk	Chalk, white, pugged up	0.5+	15.5

Grading

	Mean for deposit *percentages*			Depth below surface (m)	*percentages*					
	Fines	Sand	Gravel		Fines	Sand			Gravel	
					$-\frac{1}{16}$	$+\frac{1}{16}-\frac{1}{4}$	$+\frac{1}{4}-1$	$+1-4$	$+4-16$	$+16$
a	5	33	62	0.5–1.5	6	4	11	13	28	38
				1.5–2.5	7	9	18	13	30	23
				2.5–3.5	5	5	12	15	32	31
				3.5–4.5	3	5	13	13	30	36
				Mean	5	6	13.5	13.5	30	32
b	6	35	59	4.5–5.5	15	9	19	14	24	19
				5.5–6.5	9	8	12	13	25	33
				6.5–7.5*	12	4	17	14	30	23
				7.5–8.5*	6	6	22	17	33	16
				8.5–9.5*	6	6	15	8	32	33
				9.5–10.5*	3	3	16	15	28	35
				10.5–11.5*	5	2	9	12	38	34
				11.5–12.5*	4	3	19	16	34	24
				12.5–13.5*	3	2	17	16	32	30
				13.5–15.0*	3	1	11	20	33	32
				Mean	6	4	16	15	31	28

Composition

	Depth below surface (m)	Percentages by weight in + 4–16 mm Fraction					
		Chalk	Flint	Quartz and Quartzite	Fossil debris	Sandstone	Others
a	0.5–1.5	–	83	14	–	2	1
	1.5–2.5	–	70	19	–	9	2
	2.5–3.5	–	82	$5\frac{1}{2}$	–	7	$5\frac{1}{2}$
	3.5–4.5	–	79	11	–	2	8
	Mean	–	79	12	–	5	4
b	4.5–5.5	3	72	17	–	6	2
	5.5–6.5	$2\frac{1}{2}$	$77\frac{1}{2}$	9	–	6	5
	6.5–7.5	10	77	7	–	3	3
	7.5–8.5	10	68	14	–	3	5
	8.5–9.5	10	71	11	–	3	5
	9.5–10.5	13	70	13	1	–	3
	10.5–11.5	17	63	10	–	4	6
	11.5–12.5	17	62	10	–	4	7
	12.5–13.5	20	55	15	1	5	4
	13.5–15.0	11	$70\frac{1}{2}$	9	1	$2\frac{1}{2}$	6
	Mean	11	69	11	–	4	5

TL 41 NW 15 **4281 1531** **Marshland Wood, Widford** **Block B**

Surface level +81.7 m
Water not struck
Shell 152 mm diameter
October 1975

Waste 18.6 m+

Log

Geological classification	*Lithology*	*Thickness* m	*Depth* m
	Soil	0.1	0.1
Boulder Clay	Clay, pebbly, brown, soft, sticky	0.4	0.5
	Clay, chalky, pebbly, silty, brown mottled grey, with flint and quartz, soft	5.3	5.8
	Clay, silty, sandy, chalky, pebbly, olive green, firm	0.9	6.7
	Clay, silty, chalky, blue grey, firm	0.7	7.4
	Clay, very silty, chalky, grey, soft	0.2	7.6
	Clay, chalky, silty, blue grey, becomes black with rare chalk at base, stiff to hard	6.6	14.2
	Silt, clayey, sandy, laminated, olive green, soft	2.0	16.2
	Clay, very chalky, silty, blue black, with a trace of fossil debris, very hard	2.4+	18.6

TL 41 NW 16 **4312 1983** **The Palace, Much Hadham** **Block E**

Surface level +55.7 m
Water struck at +53.7 m
Shell (modified) 152 mm diameter
February 1976

Overburden 2.0 m
Mineral 1.2 m
Bedrock 0.4 m +

Log

Geological classification	*Lithology*	*Thickness* m	*Depth* m
	Soil	0.2	0.2
Alluvium	Silt, clayey, sandy, pebbly, grey brown becoming darker with depth, with peaty organic material, soft to very soft	1.8	2.0
	Gravel Sand: medium with coarse and some fine, predominantly angular flint, with some rounded fine quartz and quartzite Gravel: fine with coarse, subrounded to rounded flint, with rounded quartz and angular flint, and a trace of subangular sandstone	1.2	3.2
Upper Chalk	Chalk, putty, white to cream, soft	0.4+	3.6

Grading

Mean for deposit *percentages*			Depth below surface (m)	*percentages*					
Fines	Sand	Gravel		Fines	Sand			Gravel	
				$-\frac{1}{16}$	$+\frac{1}{16}-\frac{1}{4}$	$+\frac{1}{4}-1$	+1–4	+4–16	+16
3	29	68	2.0–3.2*	3	5	14	10	39	29

TL 41 NW 17 **4338 1875** **Danebridge Lane, Much Hadham** **Block A**

Surface level +59.4 m
Water not struck
Shell (modified) 152 mm diameter
February 1976

Waste 4.1 m
Bedrock 0.3 m+

Log

Geological classification	*Lithology*	*Thickness* m	*Depth* m
	Peat, pebbly, black, friable	0.5	0.5
Boulder Clay	Clay, silty, sandy, chalky, pebbly, brown, soft	0.4	0.9
	'Very clayey' pebbly sand: medium and fine quartz sand, with fine gravel of angular to well rounded flints	0.2	1.1
	Clay, silty, faintly laminated, yellow brown, with a trace of chalk, soft	2.1	3.2
Glacial Sand and Gravel	'Very clayey' sandy gravel Sand: fine to medium with some coarse, predominantly quartz, with angular coarse flint Gravel: fine with some coarse angular flints, with some well rounded flint and quartzite	0.9	4.1
Upper Chalk	Chalk, very flinty, large cobbles at top, dry	0.3+	4.4

TL 41 NW 18 4370 1818 Uffords Farm, Much Hadham **Block A**

Surface level +87.6 m
Water not struck
Shell (modified) 152 mm diameter
January 1976

Overburden 15.5 m
Mineral 6.1 m
Bedrock 1.0 m+

Log

Geological classification	*Lithology*	*Thickness* m	*Depth* m
	Soil and subsoil	0.7	0.7
Boulder Clay	Clay, chalky to very chalky, flinty, brown, soft	2.0	2.7
	Clay, sandy, chalky, yellow brown, soft	0.3	3.0
	Clay, chalky, silty, grey brown, soft becoming stiff	2.0	5.0
	Clay, chalky, grey blue becoming dark grey blue, with rare flint pebbles, and a chalky fine sand lens at 7.8 m, stiff	6.2	11.2
	Clay, silty, small chalk pellets, flinty, quartz pebbles, grey black, very stiff	0.6	11.8
	Clay, chalky, pebbly, brown, black, firm	3.4	15.2
	Clay, chalky, sandy, very pebbly, yellow brown, firm	0.3	15.5
Glacial Sand and Gravel	Sandy gravel, pebbly sand in upper 3.0 m Sand: medium with some fine and coarse, predominantly quartz, with some angular coarse and medium flint, yellow Gravel: fine with some coarse, angular to subrounded flint, with some rounded quartz and quartzite, and a trace of sandstone, very rare chalk	6.1	21.6
Woolwich and Reading Beds	Clay, light blue mottled red, brown, and green, stiff to hard, waxy	1.0+	22.6

Grading

Mean for deposit *percentages*			Depth below surface (m)	*percentages*					
Fines	Sand	Gravel		Fines	Sand			Gravel	
				$-\frac{1}{16}$	$+\frac{1}{16}-\frac{1}{4}$	$+\frac{1}{4}-1$	+1–4	+4–16	+16
5	62	33	15.5–16.5	8	36	24	6	17	9
			16.5–17.5	4	20	61	3	10	2
			17.5–18.5	3	8	80	2	5	2
			18.5–19.5	4	9	20	13	31	23
			19.5–20.5	5	6	22	15	35	17
			20.5–21.6	5	6	27	16	31	15
			Mean	5	14	39	9	22	11

Composition

Depth below surface (m)	Percentages by weight in + 4–16 mm fraction					
	Chalk	Flint	Quartz and Quartzite	Fossil debris	Sandstone	Others
15.5–16.5	3	51	20	–	17	9
16.5–17.5	–	52	31	–	8½	8½
17.5–18.5	–	59	31	–	6	4
18.5–19.5	–	51	38	–	9	2
19.5–20.5	–	56	35	–	2	7
20.5–21.6	–	52	43	–	4	1
Mean	½	53	33	–	8	5½

TL 41 NW 19 4319 1751 Hadham Towers, Much Hadham **Block A**

Surface level +77.6 m
Water struck at +74.2 m
Shell (modified) 152 mm diameter
January 1976

Overburden 3.4 m
Mineral 1.2 m
Waste 4.9 m
Mineral 4.5 m
Bedrock 4.0 m+

Log

Geological classification	*Lithology*	*Thickness* m	*Depth* m
	Soil	0.3	0.3
Boulder Clay	Clay, pebbly, silty, brown, soft	0.3	0.6
	Clay, pebbly, chalky, silty, brown, soft	1.2	1.8
	Clay, pebbly, silty, sandy, brown, with rare chalk, soft	1.6	3.4
Glacial Sand and Gravel	**a** 'Clayey' gravel Sand: medium and coarse with some fine, predominantly quartz, with angular coarse flint brown Gravel: coarse and fine, angular to rounded flint, with some rounded quartz and quartzite, and a trace of sandstone and chalk	1.2	4.6
Boulder Clay	Clay, very sandy, very chalky, pebbly, yellow, firm	2.2	6.8
	Clay, chalky, pebbly, silty, blue brown becomes olive brown, firm	2.7	9.5
Glacial Sand and Gravel	**b** Sandy gravel, pebbly sand between 11.4 and 12.4 m Sand: medium with fine and some coarse, predominantly quartz, with some angular coarse flint Gravel: fine with coarse and a trace of cobble, angular to rounded flint, and rounded quartz and quartzite, with a trace of sandstone	4.5	14.0
London Clay (Basement Bed)	Silt, very clayey, fine sandy, laminated, yellow brown mottled grey, soft	1.5	15.5
	Sand, fine, silty, clayey, olive green, with shell debris, soft	0.3	15.8
	Sand, fine, silty, becomes very clayey and glauconitic, olive grey, with shell debris, and wood fragments, and white, brown and black well rounded flints at base	1.8	17.6
Woolwich and Reading Beds	Clay, blue and brown mottled becomes brown mottled, red, green and blue, very hard, waxy	0.4+	18.0

Grading

	Mean for deposit *percentages*			Depth below surface (m)	*percentages*						
	Fines	Sand	Gravel		Fines	Sand			Gravel		
					$-\frac{1}{16}$	$+\frac{1}{16}-\frac{1}{4}$	$+\frac{1}{4}-1$	+1–4	+4–16	+16–64	+64
a	12	32	56	3.4–4.6*	12	5	15	12	27	29	–
b	8	50	42	9.5–10.5	10	6	17	17	33	17	–
				10.5–11.4	9	7	23	13	30	18	–
				11.4–12.4	7	48	38	1	3	3	–
				12.4–13.4	5	15	16	8	24	23	9
				13.4–14.0	6	3	23	12	26	28	–
				Mean	8	17	23	10	23	17	2

Composition

	Depth below surface (m)	Percentages by weight in +4–16 mm fraction					
		Chalk	Flint	Quartz and Quartzite	Fossil debris	Sandstone	Others
a	3.4–4.6	–	73	20	–	6	1
b	9.5–10.5	–	55	34	–	4	7
	10.5–11.4	–	55	38	–	5	2
	11.4–12.4	–	65	29	–	3	3
	12.4–13.4	–	48	38	–	10	4
	13.4–14.0	–	45	43	–	9	3
	Mean	–	54	36	–	6	4

TL 41 NW 20 4378 1675 Mingers, Much Hadham **Block A**

Surface level +81.7 m
Water not struck
Shell (modified) 152 mm diameter
January 1976

Waste 22.7 m
Bedrock 1.5 m+

Log

Geological classification	*Lithology*	*Thickness* m	*Depth* m
	Soil	0.4	0.4
Boulder Clay	Clay, chalky to very chalky, silty, brown, with some flint pebbles, soft	4.8	5.2
	Clay, silty, chalky, dark grey, with some flints and a trace of vein quartz, black shales and grey limestone, stiff	2.5	7.7
	Clay, very sandy, chalky, grey, firm	0.2	7.9
	Clay, chalky, dark blue grey, with flint, very stiff	3.1	11.0
	Clay, small chalk pellets, dark grey black, hard	1.5	12.5
	Clay, chalky, silty, grey brown, with flint pebbles, stiff	3.1	15.6
	Clay, chalky, grey blue, with flint, quartz and sandstone, stiff	1.4	17.0
	Sand, very chalky, silty and clayey, with flint pebbles, brown	0.6	17.6
	Clay, chalky, pebbly, sandy, brown, firm	1.1	18.7
Glacial Sand and Gravel	Gravel, clayey at top Sand: medium with coarse and some fine, predominantly angular coarse and medium flint, with some medium and fine rounded quartz Gravel: fine with coarse, angular to rounded flint, with some rounded to subrounded quartz and quartzite, and a trace of sandstone	4.0	22.7
Woolwich and Reading Beds	Clay, brown mottled red, hard, waxy	0.7	23.4
	Clay, mottled red, blue, and green, hard, waxy	0.8+	24.2

Grading

Mean for deposit *percentages*			Depth below surface (m)	*percentages*					
Fines	Sand	Gravel		Fines	Sand			Gravel	
				$-\frac{1}{16}$	$+\frac{1}{16}-\frac{1}{4}$	$+\frac{1}{4}-1$	$+1-4$	$+4-16$	$+16$
7	43	50	18.7–19.7	11	16	20	13	27	13
			19.7–20.7	4	8	14	17	31	26
			20.7–21.7	6	10	20	14	28	22
			21.7–22.7	6	7	16	16	35	20
			Mean	7	10	18	15	30	20

Composition

Depth below surface (m)	Percentages by weight in + 4–16 mm fraction					
	Chalk	Flint	Quartz and Quartzite	Fossil debris	Sandstone	Others
18.7–19.7	–	69	19	–	8	4
19.7–20.7	–	64	25	–	7	4
20.7–21.7	–	71	20	–	6	3
21.7–22.7	–	62	26	–	7	5
Mean	–	66½	22½	–	7	4

TL 41 NW 21 **4381 1604** **Turtle Farm, Widford** **Block C**

Surface level +72.2 m
Water not struck
Shell (modified) 152 mm diameter
January 1976

Overburden 9.3 m
Mineral 4.8 m
Bedrock 2.7 m+

Log

Geological classification	*Lithology*	*Thickness* m	*Depth* m
	Soil	0.3	0.3
Boulder Clay	Clay, sandy, brown, some flints, soft	0.4	0.7
	Clay, sandy, chalky, brown mottled yellow brown, some flints, firm	2.1	2.8
	Clay, chalky, silty, grey becomes olive grey, with a trace of flint, black shale and fossil debris, stiff	6.4	9.2
	Clay, brown, chalky, sandy, soft	0.1	9.3
Glacial Sand and Gravel	Sandy gravel, very sandy at top, gravel at base Sand: medium with coarse and fine, predominantly rounded quartz and angular flint, with some sandstone, yellow brown Gravel: fine with coarse, subangular to rounded flint, with some rounded quartz and quartzite, and a trace of sandstone	4.8	14.1
London Clay	Clay, very sandy, coarsely laminated, brown, soft	0.4	14.5
	Silt, clayey, laminated, light grey blue, firm	0.1	14.6
	Sand, very clayey, and silty, drab olive green, with glauconite, shell debris and pyrite nodules, soft	2.2+	16.8

Grading

Mean for deposit *percentages*			Depth below surface (m)	*percentages*					
Fines	Sand	Gravel		Fines	Sand			Gravel	
				$-\frac{1}{16}$	$+\frac{1}{16}-\frac{1}{4}$	$+\frac{1}{4}-1$	+1–4	+4–16	+16
6	48	46	9.3–10.3	7	11	52	7	16	7
			10.3–11.3	8	12	27	11	29	13
			11.3–12.3	5	10	21	12	36	16
			12.3–13.3	6	5	17	16	36	20
			13.3–14.1	4	5	13	18	29	31
			Mean	6	9	26	13	29	17

Composition

Depth below surface (m)	Percentages by weight in +4–16 mm fraction				
	Chalk	Flint	Quartz and Quartzite	Fossil debris	Sandstone	Others
9.3–10.3	–	70	16	–	10	4
10.3–11.3	–	73	14	–	8	5
11.3–12.3	–	66	27	–	4	3
12.3–13.3	–	77	15	–	4	4
13.3–14.1	–	62	19	–	6	13
Mean	–	70	18	–	6	6

TL 41 NW 22 4436 1978 Dane Bridge, Much Hadham **Block A**

Surface level +77.2 m
Water not struck
Shell (modified) 152 mm diameter
February–March 1976

Overburden 2.2 m
Mineral 9.9 m
Bedrock 3.4 m+

Log

Geological classification	*Lithology*	*Thickness* m	*Depth* m
	Soil	0.1	0.1
Head	Clay, very pebbly, orange brown, soft	1.4	1.5
	Clay, very sandy, very pebbly, chalky, brown	0.7	2.2
Glacial Sand and Gravel	Gravel, clayey at top, with a band of brown chalky silty clay between 7.1 and 7.4 m Sand: medium with some coarse and fine, predominantly quartz, with angular coarse and medium flint Gravel: fine and coarse, angular to subrounded flint, with rounded quartz and quartzite, and some rounded chalk and sandstone, with a trace of fossil debris	9.9	12.1
Woolwich and Reading Beds	Sand, fine, increasingly clayey, very silty, yellow green, firm	3.2	15.3
	Clay, sandy, pebbly, orange brown, with well rounded black flints	0.1	15.4
Upper Chalk	Chalk, pugged up, white to cream, very soft	0.1+	15.5

Grading

Mean for deposit *percentages*			Depth below surface (m)	*percentages*					
Fines	Sand	Gravel		Fines	Sand			Gravel	
				$-\frac{1}{16}$	$+\frac{1}{16}-\frac{1}{4}$	$+\frac{1}{4}-1$	+1–4	+4–16	+16
8	41	51	2.2–3.2	11	8	31	14	26	10
			3.2–4.2	9	8	24	12	25	22
			4.2–5.2	10	8	23	14	26	19
			5.2–6.2	12	8	15	13	24	28
			6.2–7.1	8	12	20	9	25	26
			7.4–8.4	6	8	21	9	26	30
			8.4–9.4	4	6	15	15	31	29
			9.4–10.4	6	5	14	16	26	33
			10.4–11.4	6	6	19	14	28	27
			11.4–12.1	5	5	25	16	27	22
			Mean	8	7	21	13	26	25

Composition

Depth below surface (m)	Percentages by weight in +4–16 mm fraction					
	Chalk	Flint	Quartz and Quartzite	Fossil debris	Sandstone	Others
2.2–3.2	27	51	11	1	7	3
3.2–4.2	28	49	12	1	9	1
4.2–5.2	25	53	12	1	7	2
5.2–6.2	16	59	8	4	3	10
6.2–7.1	3	54	30	1	7	5
7.4–8.4	–	56	25	–	15	4
8.4–9.4	–	67	24	–	6	3
9.4–10.4	–	75	19	–	4	2
10.4–11.4	–	58	27	–	11	4
11.4–12.1	–	49	36	–	13	2
Mean	10	57	20	1	8	4

TL 41 NW 23 **4480 1867** **Green Tye, Much Hadham** **Block A**

Surface level +89.8 m
Water struck at +78.3 m
Shell (modified) 152 mm diameter
February 1976

Waste 22.0 m
Bedrock 2.4 m+

Log

Geological classification	*Lithology*	*Thickness* m	*Depth* m
	Soil	0.2	0.2
Boulder Clay	Clay, very chalky, yellow brown, with a trace of flint and quartz pebbles, soft to very soft	5.7	5.9
	Clay, chalky, silty, grey blue becomes dark grey with depth, with some flint and quartz pebbles, stiff	5.1	11.0
	Clay, chalky, dark brown, with pebbles of flint and quartz, stiff	5.0	16.0
	Clay, chalky, dark grey, rare flints, stiff	1.0	17.0
	Clay, very sandy, very pebbly, brown becomes yellow brown, firm	0.4	17.4
Glacial Sand and Gravel	Gravel, sand band between 19.4 and 19.7 m Sand: medium with some coarse and fine, predominantly quartz, with coarse and medium angular flint, orange Gravel: fine with coarse, angular to subrounded flint, with subrounded to rounded quartz and quartzite, with a trace of sandstone	4.6	22.0
London Clay	Clayey silt, sandy, laminated, brown, soft	1.1	23.1
	Silt, clayey, fine sandy, olive green, very soft	1.3+	24.4

Grading

Mean for deposit *percentages*			Depth below surface (m)	*percentages*					
Fines	Sand	Gravel		Fines	Sand			Gravel	
				$-\frac{1}{16}$	$+\frac{1}{16}-\frac{1}{4}$	$+\frac{1}{4}-1$	+1–4	+4–16	+16
8	43	49	17.4–18.4	10	7	23	9	29	22
			18.4–19.4	8	10	34	9	25	14
			19.4–19.9	9	13	21	8	33	16
			19.9–20.9	8	8	23	14	28	19
			20.9–22.0	4	3	20	14	28	31
			Mean	8	8	24	11	28	21

Composition

Depth below surface (m)	Percentages by weight in + 4–16 mm fraction					
	Chalk	Flint	Quartz and Quartzite	Fossil debris	Sandstone	Others
17.4–18.4	–	68	24	–	7	1
18.4–19.4	–	59	30	–	10	1
19.4–19.9	–	55	34	–	7	4
19.9–20.9	–	72	21	–	6	1
20.9–22.0	–	69	19	–	7	5
Mean	–	66	25	–	7	2

TL 41 NW 24 **4440 1775** **Bucklers Hall Farm, Much Hadham** **Block A**

Surface level +79.4 m
Water struck at +76.6 m
Shell (modified) 152 mm diameter
January 1976

Overburden 15.0 m
Mineral 6.8 m
Waste 0.7 m
Bedrock 0.9 m+

Log

Geological classification	*Lithology*	*Thickness* m	*Depth* m
	Soil and subsoil	0.8	0.8
Boulder Clay	Clay, chalky, rare flints, yellow brown becomes grey, soft	2.0	2.8
	Clay, chalky, sandy, pebbly, yellow, very soft	0.3	3.1
	Clay, chalky, rare pebbles, brown, soft	0.7	3.8
	Clay, chalky, silty, blue grey becoming dark blue grey, with pebbles of flint and quartz, and a trace of black shale and fossil fragments, stiff	3.5	7.3
	Clay, small chalk pellets, blue black, stiff to hard	2.2	9.5
	Clay, chalky, silty, very dark brown, with brown flints, stiff	1.5	11.0
	Silts and clays, laminated, yellow, with some laminae of fine chalky sand becoming more frequent until at base laminated fine sands and silt. Laminae up to 30 cm	4.0	15.0
Glacial Sand and Gravel	'Clayey' sandy gravel, very clayey sand in upper 2.5 m becomes progressively more gravelly until gravel at base Sand: fine and medium with some coarse, predominantly rounded quartz, with some coarse angular flint, and a trace of rounded chalk, pale yellow Gravel: fine with coarse, subangular to rounded flint, with some rounded quartz, and a trace of sandstone	6.8	21.8
Boulder Clay	Clay, chalky, silty, very sandy and pebbly, brown, stiff	0.7	22.5
Woolwich and Reading Beds	Clay, silty, mottled blue, grey, brown and green, very stiff, waxy	0.9+	23.4

Grading

Mean for deposit *percentages*			Depth below surface (m)	*percentages*					
Fines	Sand	Gravel		Fines	Sand			Gravel	
				$-\frac{1}{16}$	$+\frac{1}{16}-\frac{1}{4}$	$+\frac{1}{4}-1$	+1–4	+4–16	+16
15	56	29	15.0–16.0	42	56	1	1	–	–
			16.0–17.0	32	60	7	1	–	–
			17.0–17.5	18	50	29	1	2	–
			17.5–18.5	5	11	57	9	13	5
			18.5–19.5	3	5	33	18	29	12
			19.5–20.5	4	4	16	17	36	23
			20.5–21.8	4	5	13	16	35	27
			Mean	15	25	21	10	18	11

Composition

Depth below surface (m)	Percentages by weight in + 4–16 mm fraction					
	Chalk	Flint	Quartz and Quartzite	Fossil debris	Sandstone	Others
15.0–16.0	Sand – no sample					
16.0–17.0	Sand – no sample					
17.0–17.5	Sand – very small sample					
17.5–18.5	–	64	27	–	4	5
18.5–19.5	–	69	25	–	5	1
19.5–20.5	–	62	28	–	5	5
20.5–21.8	–	61	31	–	5	3
Mean	–	64	28	–	5	3

TL 41 NW 25 4465 1670 Old Park Farm, Much Hadham **Block C**

Surface level +73.9 m
Water not struck
Shell (modified) 152 mm diameter
February 1976

Overburden 10.5 m
Mineral 4.1 m
Waste 0.8 m
Bedrock 1.5 m+

Log

Geological classification	*Lithology*	*Thickness* m	*Depth* m
	Soil	0.2	0.2
Boulder Clay	Clay, silty, yellow brown, with a trace of flint, stiff	0.6	0.8
	Clay, chalky, silty, yellow brown becoming brown streaked grey, flints, soft	1.7	2.5
	Clay, pebbly, chalky, silty, grey, soft	0.4	2.9
	Clay, chalky, flinty, dark grey blue, stiff to hard	2.5	5.4
	Clay, chalky, grey brown, with flint and quartz, stiff	3.4	8.8
	Clay, chalky, silty, grey blue, with pebbles of flint, quartz sandstone and rare black shale, stiff	0.7	9.5
	Clay, chalky, sandy at base, dark grey brown, with flint and quartz	1.0	10.5
Glacial Sand and Gravel	Sandy gravel, very sandy at top more gravelly with depth Sand: medium with some fine and coarse, predominantly rounded quartz, with coarse and medium angular flint Gravel: fine with some coarse and a trace of cobble, subangular to rounded flint, with some rounded quartz and quartzite, and a trace of sandstone	4.1	14.6
Boulder Clay	Clay, very sandy, silty, brown streaked grey, with a trace of quartz pebbles, firm	0.7	15.3
	Mudstone, rounded grey boulder	0.1	15.4
London Clay	Silt, very clayey, sandy, brown streaked grey, with carbonaceous fragments, firm	0.5	15.9
	Silt, very clayey, sandy, dark olive green, with glauconite and occasional pyrite nodules, firm	1.0+	16.9

Grading

Mean for deposit *percentages*			Depth below surface (m)	*percentages*						
Fines	Sand	Gravel		Fines	Sand			Gravel		
				$-\frac{1}{16}$	$+\frac{1}{16}-\frac{1}{4}$	$+\frac{1}{4}-1$	+1–4	+4–16	+16–64	+64
6	58	36	10.5–11.5	7	18	34	9	22	10	–
			11.5–12.5	4	12	51	10	17	6	–
			12.5–13.5	5	8	35	10	23	9	10
			13.5–14.6	7	9	20	17	28	19	–
			Mean	6	12	35	11	23	11	2

Composition

Depth below surface (m)	Percentages by weight in +4–16 mm fraction					
	Chalk	Flint	Quartz and Quartzite	Fossil debris	Sandstone	Others
10.5–11.5	–	76	16	–	5	3
11.5–12.5	–	73	23	–	2	2
12.5–13.5	–	76	20	–	2	2
13.5–14.6	–	74	20	–	4	2
Mean	–	76	19	–	3	2

TL 41 NW 26 4498 1579 Stonards, High Wych **Block C**

Surface level +76.2 m
Water struck at +57.6 m
Shell (modified) 152 mm diameter
April 1976

Overburden 13.7 m
Mineral 6.1 m
Bedrock 2.2 m+

Log

Geological classification	*Lithology*	*Thickness* m	*Depth* m
	Soil	0.3	0.3
Boulder Clay	Clay, silty, flinty, brown, stiff	0.7	1.0
	Clay, chalky, silty, yellow brown becomes brown mottled blue grey, soft	2.4	3.4
	Clay, chalky, silty, grey, soft becoming stiff	5.0	8.4
	Clay, chalky, silty, flinty, dark blue grey, stiff	3.9	12.3
	Clay, chalky, silty, very pebbly and chalkless at base, grey brown, firm	1.4	13.7
Glacial Sand and Gravel	Gravel, sandy at top becomes coarse with depth Sand: medium and coarse with some fine, predominantly rounded quartz, with coarse angular flint, yellow brown Gravel: coarse and fine, angular to rounded flint, with some rounded quartz, quartzite and sandstone	6.1	19.8
London Clay	Clay, sandy, pebbly, micaceous, laminated, brown, stiff	0.3	20.1
	Silt, clayey, micaceous, laminated in parts, dark olive grey, stiff	1.9+	22.0

Grading

Mean for deposit *percentages*			Depth below surface (m)	*percentages*					
Fines	Sand	Gravel		Fines	Sand			Gravel	
				$-\frac{1}{16}$	$+\frac{1}{16}-\frac{1}{4}$	$+\frac{1}{4}-1$	+1–4	+4–16	+16
5	38	57	13.7–14.7	8	12	24	14	28	14
			14.7–15.7	7	9	18	15	32	19
			15.7–16.7	7	8	18	17	31	19
			16.7–17.7	4	6	13	13	23	41
			17.7–18.6	3	4	12	19	37	25
			18.6–19.8*	1	4	12	10	21	52
			Mean	5	7	16	15	28	29

Composition

Depth below surface (m)	Percentages by weight in + 4–16 mm Fraction					
	Chalk	Flint	Quartz and Quartzite	Fossil debris	Sandstone	Others
13.7–14.7	–	71	20	–	5	4
14.7–15.7	–	68	21	–	10	1
15.7–16.7	–	73	18	–	5	4
16.7–17.7	–	73	18	–	8	1
17.7–18.6	–	78	12	–	6	4
18.6–19.8	–	70	19	–	9	2
Mean	–	72	18	–	7	3

TL 41 NW 27 4408 1557 Fiddlers Brook, Much Hadham **Block C**

Surface level +72.2 m
Water struck at +56.7 m
Shell (modified) 152 mm diameter
February 1976

Waste 17.6 m
Bedrock 1.1 m+

Log

Geological classification	*Lithology*	*Thickness* m	*Depth* m
	Soil	0.3	0.3
Boulder Clay	Clay, chalky, silty, rare flint, yellow brown, soft	2.8	3.1
	Clay, chalky, silty, rarely pebbly, brown, soft to stiff	1.3	4.4
	Clay, very chalky, silty, some flints, grey blue, stiff	2.4	6.8
	Clay, silty, chalky, flint and quartz pebbles, dark grey black, hard	2.1	8.9
	Clay, chalky to very chalky, silty, pebbly, light grey, stiff	1.5	10.4
	Clay, chalky, pebbly, very flinty, brown, firm	3.0	13.4
	Clay, very pebbly, very sandy, yellow brown, firm	0.1	13.5
Glacial Sand and Gravel	Gravel Sand: medium with coarse and some fine, predominantly angular coarse and medium flint, with fine rounded quartz, brown Gravel: fine and coarse, angular to subrounded flint, with rounded quartz and quartzite, and a trace of sandstone	4.1	17.6
London Clay	Clay, very silty, sandy, yellow brown, firm	0.2	17.8
	Silt, very clayey, fine sandy, drab olive green, with decalcified shell debris, soft	0.9+	18.7

Grading

Mean for deposit *percentages*			Depth below surface (m)	*percentages*					
Fines	Sand	Gravel		Fines	Sand			Gravel	
				$-\frac{1}{16}$	$+\frac{1}{16}-\frac{1}{4}$	$+\frac{1}{4}-1$	+1–4	+4–16	+16
4	44	52	13.5–14.5	7	10	20	15	29	19
			14.5–15.5	7	9	16	16	30	22
			15.5–16.5*	1	7	18	15	25	34
			16.5–17.6*	2	14	22	12	21	29
			Mean	4	10	19	15	26	26

Composition

Depth below surface (m)	Percentages by weight in + 4–16 mm fraction					
	Chalk	Flint	Quartz and Quartzite	Fossil debris	Sandstone	Others
13.5–14.5	–	66	29½	–	3½	1
14.5–15.5	–	76	21	–	2½	½
15.5–16.5	–	65	31	–	3	1
16.5–17.6	–	65	26	–	6	3
Mean	–	68	27	–	4	1

TL 41 NE 13 **4570 1976** **Jobbers Wood, Much Hadham** **Block A**

Surface level +93.3 m
Water struck at +85.3 m
Shell (modified) 152 mm diameter
March 1976

Waste 18.6 m+

Log

Geological classification	*Lithology*	*Thickness* m	*Depth* m
	Soil	0.2	0.2
Head	Clay, sandy, silty, yellow brown, with a trace of flint pebbles, soft	0.8	1.0
Boulder Clay	Clay, very chalky, silty, yellow brown, with chalk and flint cobbles, firm	1.8	2.8
	Clay, chalky, silty, grey, with some flints, firm to stiff	5.2	8.0
	Silt and chalk sand, pale grey, with fine chalk gravel, soft	0.9	8.9
	Clay, chalky, silty, grey becoming dark blue grey, with some flints, stiff	9.7+	18.6

TL 41 NE 14 4532 1891 Warren Farm, Much Hadham **Block A**

Surface level +88.5 m
Water struck at +82.6 m and +78.5 m
Shell (modified) 152 mm diameter
February 1976

Waste 21.0 m
Bedrock 1.0 m+

Log

Geological classification	*Lithology*	*Thickness* m	*Depth* m
	Made ground	1.1	1.1
Boulder Clay	Clay, very chalky, pebbly, brown mottled grey, soft	4.7	5.8
	Sand, very clayey, chalky, pebbly, brown	0.3	6.1
	Clay, very chalky, silty, grey blue, soft becoming stiff	7.9	14.0
	Clay, chalky, silty, dark grey, with oolitic limestone boulder between 14.5 m and 14.8 m, stiff	3.4	17.4
	Clay, silty, many sand sized chalk pellets, dark grey, stiff	0.5	17.9
	Clay, sandy, silty, pebbly, yellow brown, firm	0.2	18.1
Glacial Sand and Gravel	'Clayey' sandy gravel, less 'clayey' with depth Sand: medium with fine and coarse, predominantly quartz, with some angular coarse flint, orange brown Gravel: fine with coarse, subrounded to angular flint, with sub-rounded to rounded quartz and quartzite, and some rounded sandstone	2.7	20.8
	Clay, very sandy, silty, brown, soft	0.2	21.0
Woolwich and Reading Beds	Clay, silty, mottled red and brown streaked green and light blue, with a trace of fine gravel of reddened flints, stiff becoming hard, waxy	1.0+	22.0

Grading

Mean for deposit *percentages*			Depth below surface (m)	*percentages*					
Fines	Sand	Gravel		Fines	Sand			Gravel	
				$-\frac{1}{16}$	$+\frac{1}{16}-\frac{1}{4}$	$+\frac{1}{4}-1$	+ 1–4	+ 4–16	+ 16
13	46	41	18.1–19.1	14	12	39	6	16	13
			19.1–20.1	18	9	18	11	21	23
			20.1–20.8	5	10	18	13	36	18
			Mean	13	10	26	10	23	18

Composition

Depth below surface (m)	Percentages by weight in + 4–16 mm fraction					
	Chalk	Flint	Quartz and Quartzite	Fossil debris	Sandstone	Others
18.1–19.1	–	60	30	–	7	3
19.1–20.1	–	67	27½	–	4½	1
20.1–20.8	–	66	29	–	2	3
Mean	–	64	29	–	5	2

TL 41 NE 15 4589 1850 Mathams Wood, High Wych **Block A**

Surface level +93.5 m
Water struck at +85.0 m
Shell (modified) 152 mm diameter
April 1976

Waste 19.7 m+

Log

Geological classification	*Lithology*	*Thickness* m	*Depth* m
	Soil	0.3	0.3
Boulder Clay	Clay, chalky, flinty, silty, brown, soft	1.1	1.4
	Clay, silty, chalky, brown mottled blue grey, soft	3.8	5.2
	Clay, silty, chalky, dark blue grey, firm	0.5	5.7
	Clay, chalky to very chalky, silty, light brown, soft	0.9	6.6
	Clay, silty, chalky, light blue grey, soft	1.1	7.7
	Clay, fine sandy, silty, chalky to very chalky, yellow brown, very soft	1.9	9.6
	Clay, silty, chalky, dark blue grey, with some flint and grey limestone, firm becoming stiff	8.2	17.8
	Sand, fine and medium, 'clayey', chalky, dark brown, soft	0.2	18.0
	Clay, silty, chalky, brown grey, soft	0.6	18.6
	Sand, fine and medium, 'clayey', chalky, dark brown, soft	0.2	18.8
	Clay, very chalky, sandy, silty, brown, soft	0.2	19.0
	Clay, chalky, silty, brown mottled blue grey, firm	0.7+	19.7

TL 41 NE 16 4539 1795 Sacombs Ash, High Wych **Block A**

Surface level +85.4 m
Water struck at +63.8 m
Shell 152 mm diameter
November 1975

Waste 18.6 m+

Log

Geological classification	*Lithology*	*Thickness* m	*Depth* m
	Soil	0.2	0.2
Boulder Clay	Clay, chalky, pebbly, silty, yellow brown becomes mottled pale blue, with angular flint, rounded quartz and fossil fragments, soft	4.7	4.9
	Clay, chalky, silty, with flint, grey blue, soft	6.7	11.6
	Silt, fine, sandy, chalky, laminated with black silt bands, pale grey, soft	1.4	13.0
Glacial Sand and Gravel	Gravel Sand: fine and coarse with medium, predominantly rounded quartz and flint, with some rounded chalk and angular coarse flint Gravel: coarse with fine and cobble, angular flint and rounded chalk, with some rounded limestone and sandstone, and a trace of quartz, quartzite and fossil debris	1.0	14.0
Boulder Clay	Clay, silty, chalky, dark blue grey, with some flint and quartz pebbles, firm	4.6+	18.6

Grading

Mean for deposit *percentages*			Depth below surface (m)	*percentages*						
Fines	Sand	Gravel		Fines	Sand			Gravel		
				$-\frac{1}{16}$	$+\frac{1}{16}-\frac{1}{4}$	$+\frac{1}{4}-1$	+1–4	+4–16	+16–64	+64
12	38	50	13.0–14.0	12	16	9	13	14	26	10

Composition

Depth below surface (m)	Percentages by weight in + 4–16 mm fraction					
	Chalk	Flint	Quartz and Quartzite	Fossil debris	Sandstone	Others
13.0–14.0	35	32	3	2	5	23*

* Including limestone.

TL 41 NE 17 **4546 1527** **Hoskins Farm, High Wych** **Block C**

Surface level +77.5 m
Water struck at +74.4 m
Shell (modified) 152 mm diameter
April 1976

Waste 20.6 m+

Log

Geological classification	*Lithology*	*Thickness* m	*Depth* m
	Soil	0.2	0.2
Boulder Clay	Clay, silty, sandy, brown, soft	0.7	0.9
	Clay, silty, chalky, brown, soft	2.2	3.1
	Clay, sandy, silty, yellow brown, very stoney at base, soft	1.9	5.0
	Clay, chalky, silty, grey, firm becoming stiff	7.5	12.5
	Mudstone boulder, dark grey, very hard	0.2	12.7
	Clay, chalky, silty, dark grey, stiff	0.9	13.6
	Clay, chalky, silty, dark brown, stiff becoming firm	6.4	20.0
	Clay, pebbly, silty, sandy, brown, firm	0.1	20.1
Glacial Sand and Gravel	Gravel Sand: medium with coarse and some fine, predominantly rounded quartz, with some angular coarse flint, brown Gravel: fine with coarse, subangular to rounded flint, with rounded to subrounded quartz, quartzite and sandstone	0.5+	20.6

Grading

Mean for deposit *percentages*			Depth below surface (m)	*percentages*					
Fines	Sand	Gravel		Fines	Sand			Gravel	
				$-\frac{1}{16}$	$+\frac{1}{16}-\frac{1}{4}$	$+\frac{1}{4}-1$	+1–4	+4–16	+16–64
7	40	53	20.1–20.6	7	6	22	12	35	18

Composition

Depth below surface (m)	Percentages by weight in + 4–16 mm fraction					
	Chalk	Flint	Quartz and Quartzite	Fossil debris	Sandstone	Others
20.1–20.6	–	69	18	–	7	6

TL 41 NE 18 4667 1953 Butlers Hall Farm, Thorley **Block D**

Surface level +88.3 m
Water struck at +76.6 m
Shell (modified) 152 mm diameter
March 1976

Overburden 16.3 m
Mineral 6.7 m
Bedrock 1.5 m+

Log

Geological classification	*Lithology*	*Thickness* m	*Depth* m
	Soil	0.2	0.2
Boulder Clay	Clay, silty, sandy, chalky, orange brown, soft	0.8	1.0
	Clay, chalky, silty, fawn, soft	3.7	4.7
	Clay, chalky, silty, grey becoming blue grey, with some flint and quartz pebbles, firm to stiff	6.9	11.6
	Clay, chalky, silty, gritty, light grey, soft	0.2	11.8
	Clay, chalky, silty, blue grey, firm	3.7	15.5
	Clay, chalky, silty, pebbly, dark brown, stiff	0.5	16.0
	Clay, sandy, silty, very pebbly, yellow brown, soft to firm	0.3	16.3
Glacial Sand and Gravel	'Clayey' sandy gravel, 'very clayey' pebbly sand at top, grading to gravel Sand: medium with some fine and coarse, predominantly rounded quartz, with a trace of angular coarse flint, yellow Gravel: fine with coarse, angular to subrounded flint, with rounded quartz and quartzite, and some rounded sandstone	6.7	23.0
Woolwich and Reading Beds	Clay, brown mottled grey, waxy, with well developed laminae of light brown fine silt, firm to stiff	0.6	23.6
	Clay and sands, laminated, light brown and dark grey, waxy, soft	0.2	23.8
	Sand, fine, clayey, laminated, yellow brown, firm	0.2	24.0
	Sand, fine, laminated, white, with pellets of light grey waxy clay, firm	0.5+	24.5

Grading

Mean for deposit *percentages*			Depth below surface (m)	*percentages*					
Fines	Sand	Gravel		Fines	Sand			Gravel	
				$-\frac{1}{16}$	$+\frac{1}{16}-\frac{1}{4}$	$+\frac{1}{4}-1$	+1–4	+4–16	+16
10	50	40	16.3–17.3	30	23	25	4	11	7
			17.3–18.3	13	13	21	10	27	16
			18.3–19.3	5	34	29	6	16	10
			19.3–20.3	6	5	25	15	27	22
			20.3–21.3	4	8	30	10	28	20
			21.3–22.3	4	5	32	13	25	21
			22.3–23.0	4	4	23	14	35	20
			Mean	10	13	27	10	24	16

Composition

Depth below surface (m)	Percentages by weight in + 4–16 mm fraction					
	Chalk	Flint	Quartz and Quartzite	Fossil debris	Sandstone	Others
16.3–17.3	–	59	30	–	4	7
17.3–18.3	–	54	29	–	13	4
18.3–19.3	–	52	31	–	11	6
19.3–20.3	–	48	28	–	21	3
20.3–21.3	–	53	27	–	16	4
21.3–22.3	–	40	35	–	19	6
22.3–23.0	–	37	41	–	16	6
Mean	–	50	31	–	14	5

TL 41 NE 19 **4678 1835** **Shingle Hall, High Wych** **Block D**

Surface level +83.7 m
Water not struck
Shell 152 mm diameter
November 1975

Waste 18.6 m+

Log

Geological classification	*Lithology*	*Thickness* m	*Depth* m
	Soil	0.1	0.1
	Made ground, concrete fill	0.8	0.9
Boulder Clay	Clay, chalky, brown, stiff	4.3	5.2
	Clay, silty, chalky, grey becoming blue grey, with some flint pebbles, stiff	13.4+	18.6

TL 41 NE 20 **4684 1747** **Shingle Hall, High Wych** **Block D**

Surface level +85.0 m
Water not struck
Shell 152 mm diameter
November 1975

Waste 18.6 m+

Log

Geological classification	*Lithology*	*Thickness* m	*Depth* m
	Soil	0.2	0.2
Boulder Clay	Clay, chalky, silty, brown, stiff	2.9	3.1
	Clay, silty, chalky, grey becoming blue grey, with some flint pebbles, stiff	15.5+	18.6

TL 41 NE 21 **4614 1694** **Sweetdews, High Wych** **Block C**

Surface level +78.9 m
Water struck at +68.0 m and +56.4 m
Shell (modified) 152 mm diameter
April–May 1976

Overburden 18.0 m
Mineral 6.3 m
Bedrock 0.7 m+

Log

Geological classification	*Lithology*	*Thickness* m	*Depth* m
	Soil	0.2	0.2
Boulder Clay	Clay, silty, sandy, chalky, orange brown, soft	0.6	0.8
	Clay, silty, chalky, mottled grey and brown, soft	2.4	3.2
	Clay, chalky, silty, light grey, soft	4.0	7.2
	Clay, chalky, silty, dark blue grey, stiff	2.8	10.0
	Clay, chalky, silty, light grey, soft	0.6	10.6
	'Very clayey' sand, fine and medium, chalky, grey brown	0.3	10.9
	Clay, chalky, silty, sandy, brown, soft	0.4	11.3
	Clay, chalky, silty, blue grey, stiff	6.6	17.9
	Clay, pebbly, sandy, silty, brown, firm	0.1	18.0
Glacial Sand and Gravel	Gravel, becomes coarser and less 'clayey' with depth Sand: medium and coarse with some fine, predominantly quartz, with coarse angular flint, yellow brown Gravel: fine with coarse and a trace of cobble, subrounded to angular flint, with some subrounded to rounded quartz, quartzite and sandstone	6.3	24.3
Woolwich and Reading Beds	Clay, silty, brown, stiff, waxy	0.3	24.6
	Clay, silty, brown mottled red, green and light blue becomes grey mottled red brown and blue, stiff, waxy	0.4+	25.0

Grading

Mean for deposit *percentages*			Depth below surface (m)	*percentages*						
Fines	Sand	Gravel		Fines	Sand			Gravel		
				$-\frac{1}{16}$	$+\frac{1}{16}-\frac{1}{4}$	$+\frac{1}{4}-1$	+1–4	+4–16	+16–64	+64
4	35	61	18.0–19.0	9	6	15	15	29	26	–
			19.0–20.0	8	9	19	15	31	18	–
			20.0–21.0	4	6	20	7	32	31	–
			21.0–22.0	2	5	15	11	32	25	10
			22.0–23.0	3	6	15	16	40	20	–
			23.0–24.3	1	3	12	13	36	35	–
			Mean	4	6	16	13	33	26	2

Composition

Depth below surface (m)	Percentages by weight in + 4–16 mm fraction					
	Chalk	Flint	Quartz and Quartzite	Fossil debris	Sandstone	Others
18.0–19.0	–	69	22	–	8	1
19.0–20.0	–	71	19	–	6	4
20.0–21.0	–	75	19	–	2	4
21.0–22.0	–	72	16	–	10	2
22.0–23.0	–	79	14	–	7	–
23.0–24.3	–	59	30	–	9	2
Mean	–	70	21	–	7	2

TL 41 NE 22 4663 1660 Shingle Hall, High Wych **Block D**

Surface level +83.9 m
Water not struck
Shell 152 mm diameter
November 1975

Waste 18.6 m+

Log

Geological classification	*Lithology*	*Thickness* m	*Depth* m
	Soil	0.1	0.1
Boulder Clay	Clay, silty, pebbly, brown, soft	0.5	0.6
	Clay, chalky, silty, yellow brown becoming brown, with pockets of sandy clay, and flint pebbles, stiff	7.3	7.9
	Clay, silty, chalky, pebbly, blue grey becoming blue black, stiff becoming hard	10.7+	18.6

TL 41 NE 23 4627 1552 Gangies Hill, High Wych **Block C**

Surface level +79.1 m
Water struck at 71.9 m and 71.3 m
Shell (modified) 152 mm diameter
April 1976

Waste 23.8 m+

Log

Geological classification	*Lithology*	*Thickness* m	*Depth* m
	Soil	0.2	0.2
Boulder Clay	Clay, chalky, silty, ochre brown, soft	0.7	0.9
	Clay, very chalky, silty, light brown, firm	1.1	2.0
	Clay, chalky, silty, brown, soft	2.3	4.3
	Clay, chalky, silty, dark brown, soft	1.6	5.9
	Silt, clayey, laminated, grey, soft	1.0	6.9
	Silt, clayey, very chalky, grey, very soft	0.3	7.2
	Clay, silty, sandy, chalky, blue grey, stiff	0.9	8.1
	Chalk boulder, with bedded angular blue and black flints, white, hard	1.7	9.8
	Clay, chalky, silty, dark blue, with some flints, very stiff	9.0	18.8
	Clay, chalky, silty, mid-brown, soft	3.3	22.1
	Clay, chalky, pebbly, silty, dark blue grey, hard	1.7+	23.8

TL 41 NE 24 4768 1961 Thorley Place, Bishops Stortford **Block D**

Surface level + 87.6 m
Water not struck
Shell (modified) 152 mm diameter
March 1976

Waste 18.6 m+

Log

Geological classification	*Lithology*	*Thickness* m	*Depth* m
	Soil	0.2	0.2
Boulder Clay	Clay, very chalky, silty, yellow grey, with some flint pebbles, friable	3.6	3.8
	Silt, clayey, yellow grey, soft	0.1	3.9
	Clay, silty, chalky, dark grey, stiff	8.8	12.7
	Clay, sandy, silty, chalky, grey, soft	0.1	12.8
	Clay, chalky, dark blue grey, with some flint and black shaley mudstone, hard	5.8+	18.6

TL 41 NE 25 **4758 1828** **Thorley Hall Farm, Thorley** **Block D**

Surface level +79.0 m
Water struck at +67.7 m
Shell (modified) 152 mm diameter
March 1976

Overburden 13.9 m
Mineral 9.8 m
Bedrock 1.5 m+

Log

Geological classification	*Lithology*	*Thickness* m	*Depth* m
	Soil	0.2	0.2
Boulder Clay	Clay, silty, chalky, light brown, soft	3.5	3.7
	Clay, chalky, silty, dark grey, firm to stiff	6.6	10.3
	Clay, very silty, sandy, chalky, grey, with many wet silty, chalky, sand seams between 10.5 m and 12.0 m, soft	1.7	12.0
	Clay, chalky, silty, grey, firm becoming stiff	0.8	12.8
	Clay, chalky, brown, stiff	0.6	13.4
	Clay, very sandy, pebbly, brown, firm	0.5	13.9
Glacial Sand and Gravel	Gravel, sandy in top metre Sand: medium and coarse with some fine, predominantly quartz, with some angular medium and coarse flint, yellow brown Gravel: fine with coarse, well rounded to angular flint, with rounded quartz and quartzite, and some rounded sandstone, rare chalk	9.8	23.7
Woolwich and Reading Beds	Clay, very fine sandy, silty, green mottled red and orange, firm becoming stiff	1.5+	25.2

Grading

Mean for deposit percentages			Depth below surface (m)	percentages					
Fines	Sand	Gravel		Fines	Sand			Gravel	
				$-\frac{1}{16}$	$+\frac{1}{16}-\frac{1}{4}$	$+\frac{1}{4}-1$	+1–4	+4–16	+16–64
6	43	51	13.9–14.9	4	7	34	15	29	11
			14.9–15.9	12	9	17	13	29	20
			15.9–16.9	6	7	23	16	27	21
			16.9–17.9	8	10	27	12	30	13
			17.9–18.9	3	7	17	22	37	14
			18.9–19.9	7	8	15	19	41	10
			19.9–20.9	5	8	15	16	38	18
			20.9–21.9	4	7	15	18	35	21
			21.9–22.9	6	6	12	19	34	23
			22.9–23.7	5	6	11	16	40	22
			Mean	6	7	19	17	34	17

Composition

Depth below surface (m)	Percentages by weight in +4–16 mm fraction					
	Chalk	Flint	Quartz and Quartzite	Fossil debris	Sandstone	Others
13.9–14.9	–	59	29	–	8	4
14.9–15.9	–	61	21	–	13	5
15.9–16.9	1	72	18	–	6	3
16.9–17.9	1	60	20	–	13	6
17.9–18.9	3	58	24	–	10	5
18.9–19.9	3	65	14	–	14	3
19.9–20.9	2	63	18	–	11	6
20.9–21.9	2	60	22	–	12	4
21.9–22.9	–	63	21	–	9	7
22.9–23.7	–	64	21	–	9	6
Mean	1	62	21	–	11	5

TL 41 NE 26 **4764 1729** **Bursteads, High Wych** **Block D**

Surface level +72.3 m
Water struck at +55.3 m
Shell 152 mm diameter
October–November 1975

Overburden 11.4 m
Mineral 9.4 m
Bedrock 1.2 m+

Log

Geological classification	*Lithology*	*Thickness* m	*Depth* m
	Soil	0.3	0.3
Boulder Clay	Clay, chalky, silty, yellow brown, with some flints, soft	2.8	3.1
	Clay, chalky, light grey, very stiff	0.8	3.9
	Clay, chalky, blue grey, with some flint, quartz and quartzite pebbles, stiff	7.3	11.2
	Clay, pebbly, brown, stiff	0.2	11.4
Glacial Sand and Gravel	'Clayey' sandy gravel, fine with some medium sand in lower 1.6 m Sand: fine with medium and coarse, predominantly rounded quartz, with some medium and coarse angular flint, yellow, brown at base Gravel: fine with coarse, angular to rounded flint, with well rounded quartz and quartzite, and some rounded sandstone	9.4	20.8
London Clay	Silt, fine sandy, clayey, grey to grey green, with bands of comminuted shell debris, soft	1.2+	22.0

Grading

Mean for deposit *percentages*			Depth below surface (m)	*percentages*					
Fines	Sand	Gravel		Fines	Sand			Gravel	
				$-\frac{1}{16}$	$+\frac{1}{16}-\frac{1}{4}$	$+\frac{1}{4}-1$	+1–4	+4–16	+16–64
10	51	39	11.4–12.4	14	9	14	15	29	19
			12.4–13.4	13	11	20	15	24	17
			13.4–14.4	13	16	15	15	24	17
			14.4–15.4	17	12	18	13	26	14
			15.4–16.4	11	10	18	14	30	17
			16.4–17.4*	5	8	15	17	32	23
			17.4–19.2*	4	8	16	16	29	27
			19.2–20.8*	7	81	11	1	–	–
			Mean	10	22	16	13	23	16

Composition

Depth below surface (m)	Percentages by weight in + 4–16 mm fraction					
	Chalk	Flint	Quartz and Quartzite	Fossil debris	Sandstone	Others
11.4–12.4	–	67	21	–	4½	7½
12.4–13.4	–	66	20	–	10	4
13.4–14.4	–	65	20	–	9	6
14.4–15.4	–	70	21	–	7	2
15.4–16.4	–	69	24	–	3	4
16.4–17.4	–	64	24	–	9	3
17.4–19.2	–	67	23	–	5	5
19.2–20.8	Sand, no pebbles					
Mean	–	67	22	–	7	4

TL 41 NE 27 **4769 1625** **Tharbies Farm, High Wych** **Block D**

Surface level +74.6 m
Water not struck
Shell 152 mm diameter
October 1975

Waste 18.5 m
Bedrock 2.6 m+

Log

Geological classification	*Lithology*	*Thickness* m	*Depth* m
	Soil	0.1	0.1
Boulder Clay	Clay, chalky, brown, stiff	0.6	0.7
	Clay, silty, chalky, cream brown, with some flint pebbles, firm	2.8	3.5
	Clay, chalky, silty, dark brown, stiff	1.6	5.1
	Clay, chalky, silty, grey blue becoming dark grey, stiff becoming hard	11.5	16.6
	Clay, sandy, pebbly, brown, soft	0.2	16.8
Glacial Sand and Gravel	'Very clayey' sandy gravel Sand: medium with fine and coarse, angular flint and rounded quartz, pale brown Gravel: fine with coarse, mainly well rounded to angular flint, with well rounded quartz and quartzite, and some sandstone	1.7	18.5
London Clay (Basement Bed)	Clay, silty, sandy, brown, with poorly developed laminations, soft	0.9	19.4
	Sand, fine, silty, clayey, olive grey to dark blue, with some white shell debris, firm	1.7+	21.1

Grading

Mean for deposit *percentages*			Depth below surface (m)	*percentages*					
Fines	Sand	Gravel		Fines	Sand			Gravel	
				$-\frac{1}{16}$	$+\frac{1}{16}-\frac{1}{4}$	$+\frac{1}{4}-1$	+1–4	+4–16	+16
27	46	27	16.8–18.5	27	11	25	10	18	9

Composition

Depth below surface (m)	Percentages by weight in +4–16 mm fraction					
	Chalk	Flint	Quartz and Quartzite	Fossil debris	Sandstone	Others
16.8–18.5	–	68	23	–	5	4

TL 41 NE 28 4793 1594 Leventhorpe School, Sawbridgeworth **Block D**

Surface level +72.9 m
Water struck at +62.1 m
Shell (modified) 152 mm diameter
April 1976

Waste 17.8 m
Bedrock 1.0 m+

Log

Geological classification	*Lithology*	*Thickness* m	*Depth* m
	Soil	0.2	0.2
Boulder Clay	Clay, silty, sandy, chalky, light brown, with a trace of flint pebbles, soft	6.8	7.0
	Clay, silty, chalky, light brown, firm	1.6	8.6
	Clay, chalky, silty, olive grey, stiff	2.2	10.8
	'Clayey' sand, chalky, brown, soft	0.1	10.9
	Clay, chalky, silty, olive grey, with flint and quartz pebbles, stiff becoming hard	4.7	15.6
	Clay, pebbly, brown, firm	0.2	15.8
Glacial Sand and Gravel	'Very clayey' sandy gravel	0.1	15.9
	Clay, silty, very sandy, faintly laminated, brown, soft	0.2	16.1
	Gravel Sand: medium with some coarse and fine, predominantly quartz, with some angular coarse flint, yellow Gravel: fine and coarse, subangular to rounded flint, with quartz and quartzite, and some sandstone	1.7	17.8
London Clay	Clay, silty, sandy, brown, firm	0.4	18.2
	Clay, silty, sandy becoming very sandy, olive grey, with comminuted shell debris and pyrite nodules	0.6+	18.8

Grading

Mean for deposit *percentages*			Depth below surface (m)	*percentages*					
Fines	Sand	Gravel		Fines	Sand			Gravel	
				$-\frac{1}{16}$	$+\frac{1}{16}-\frac{1}{4}$	$+\frac{1}{4}-1$	+1–4	+4–16	+16
7	46	47	16.1–17.1	7	7	34	10	22	20
			17.1–17.8	8	9	14	14	32	23
			Mean	7	8	26	12	26	21

Composition

Depth below surface (m)	Percentages by weight in + 4–16 mm fraction					
	Chalk	Flint	Quartz and Quartzite	Fossil debris	Sandstone	Others
16.1–17.1	–	69	20	–	7	4
17.1–17.8	–	53	30	–	11	6
Mean	–	62	24	–	9	5

TL 41 NE 29 **4710 1548** **Tharbies Lodge, Sawbridgeworth** **Block D**

Surface level +68.2 m
Water not struck
Shell 152 mm diameter
October 1975

Overburden 11.4 m
Mineral 4.2 m
Waste 0.3 m
Bedrock 6.5 m+

Log

Geological classification	*Lithology*	*Thickness* m	*Depth* m
	Soil	0.3	0.3
Boulder Clay	Clay, very silty, chalky, pebbly, yellow brown, soft	3.6	3.9
	Clay, chalky, silty, blue grey becoming dark grey, with some flint and quartz pebbles, firm becoming stiff	7.0	10.9
	Clay, silty, sandy, very pebbly, dark brown, firm	0.5	11.4
Glacial Sand and Gravel	'Clayey' gravel Sand: medium and coarse with fine, predominantly quartz, with some coarse angular flints, yellow brown Gravel: fine and coarse, subrounded to angular flint, with some rounded quartz and quartzite, and a trace of sandstone	4.2	15.6
	Clay, sandy, pebbly, yellow brown, soft	0.3	15.9
London Clay (Basement Bed)	Sand, very silty, clayey, fine with some medium, lamination in parts, some shell debris, poorly developed horizon of rounded brown and black flint pebbles at base, drab olive green	5.9	21.8
Woolwich and Reading Beds	Clay, mottled blue, red and brown, stiff becoming hard, waxy	0.6+	22.4

Grading

Mean for deposit *percentages*			Depth below surface (m)	*percentages*					
Fines	Sand	Gravel		Fines	Sand			Gravel	
				$-\frac{1}{16}$	$+\frac{1}{16}-\frac{1}{4}$	$+\frac{1}{4}-1$	+1–4	+4–16	+16
11	40	49	11.4–12.4	10	7	19	12	29	23
			12.4–13.4	9	7	20	12	22	30
			13.4–14.4	11	9	19	12	25	24
			14.4–15.6	12	10	19	14	28	17
			Mean	11	8	19	13	26	23

Composition

Depth below surface (m)	Percentages by weight in +4–16 mm fraction				
	Chalk	Flint	Quartz and Quartzite	Fossil debris	Sandstone	Others
11.4–12.4	–	79	14	–	4	3
12.4–13.4	–	75	14	–	7	4
13.4–14.4	–	71	18	–	7	4
14.4–15.6	–	68	21	–	7	4
Mean	–	73	17	–	6	4

TL 41 NE 30 4829 1940 The Rectory, Bishops Stortford **Block D**

Surface level +80.0 m
Water struck at +73.7 m, +73.4 m and +73.3 m
Shell (modified) 152 mm diameter
March 1976

Overburden 10.8 m
Mineral 7.8 m
Bedrock 3.4 m+

Log

Geological classification	*Lithology*	*Thickness* m	*Depth* m
	Soil	0.2	0.2
Boulder Clay	Clay, chalky, silty, brown, soft	3.3	3.5
	Clay, chalky, silty, blue grey, firm becoming stiff	2.8	6.3
	Clay, silty, grey, with chalky fine sand bands at 6.3 m, 6.6 m and 6.7 m, soft	0.4	6.7
Glacial Sand and Gravel	**a** Sandy gravel Sand: medium with a trace of fine and coarse, predominantly quartz, with some flint and chalk, white Gravel: coarse and fine, subangular to angular flint, with chalk, and some rounded quartz, quartzite and sandstone, and a trace of fossil debris	2.0	8.7
Boulder Clay	Clay, chalky, silty, blue grey, sand bands to 9.3 m, firm	1.8	10.5
	Clay, chalky, silty, pebbly, brown, firm	0.3	10.8
Glacial Sand and Gravel	**b** Sandy gravel, 'clayey' at top becomes coarser with depth. Brown sandy silty clay band between 15.8 m and 16.0 m Sand: medium with some coarse and fine, predominantly quartz, with some angular coarse flint, yellow brown Gravel: fine with coarse, subangular to rounded flint, with rounded quartz and quartzite, and some rounded sandstone	7.8	18.6
London Clay (Basement Bed)	Clay, pebbly, sandy, brown	0.1	18.7
	Clay, sandy, silty, laminated, ochre brown, soft	0.4	19.1
	Sand, fine, very clayey, silty, olive green, with some wood fragments, soft	1.0	20.1
Woolwich and Reading Beds	Silt, sandy becoming very sandy, lime green, with claystone bands and nodules, firm	0.9	21.0
	Clay, blue and brown mottled becomes brown, with calcareous nodules, very stiff, waxy	1.0+	22.0

Grading

	Mean for deposit *percentages*			Depth below surface (m)	*percentages*					
	Fines	Sand	Gravel		Fines	Sand			Gravel	
					$-\frac{1}{16}$	$+\frac{1}{16}-\frac{1}{4}$	$+\frac{1}{4}-1$	+ 1–4	+ 4–16	+ 16
a	5	53	42	6.7–8.0*	6	10	37	6	17	24
				8.0–8.7*	4	6	34	12	27	17
				Mean	5	9	36	8	20	22
b	6	53	41	10.8–11.8	11	11	21	10	32	15
				11.8–12.8	5	8	28	11	31	17
				12.8–13.8	7	8	30	13	31	11
				13.8–14.8	4	5	37	13	24	17
				14.8–15.8	8	9	43	16	22	2
				16.0–17.0	5	10	54	12	15	4
				17.0–18.0	4	4	21	16	31	24
				18.0–18.6	6	6	19	18	31	20
				Mean	6	8	32	13	27	14

Composition

	Depth below surface (m)	Percentages by weight in + 4–16 mm fraction					
		Chalk	Flint	Quartz and Quartzite	Fossil debris	Sandstone	Others
a	6.7–8.0	18	40	24	1	6	11
	8.0–8.7	18	50	20	1	4	7
	Mean	18	43	23	1	5	10
b	10.8–11.8	–	51	34	–	9	6
	11.8–12.8	–	53	39	–	7	1
	12.8–13.8	–	58	26	–	10	6
	13.8–14.8	–	63	21	–	12	4
	14.8–15.8	–	63	23	–	9	5
	16.0–17.0	–	65	24	–	8	3
	17.0–18.0	–	79	15	–	4	2
	18.0–18.6	–	62	21	–	12	5
	Mean	–	62	26	–	9	3

TL 41 NE 31 **4842 1884** **Thorley Wood, Thorley** **Block D**

Surface level +64.4 m
Water struck at +59.6 m
Shell (modified) 152 mm diameter
March 1976

Overburden 2.3 m
Mineral 3.6 m
Bedrock 5.1 m+

Log

Geological classification	*Lithology*	*Thickness* m	*Depth* m
	Soil	0.3	0.3
Head	Clay, silty, fine sandy, brown, soft	1.5	1.8
	Clay, sandy, silty, grey, with carbonaceous partings, soft	0.5	2.3
Glacial Sand and Gravel	'Clayey' sand, coarse gravel in lower 0.1 m Sand: fine with some medium and a trace of coarse, predominantly rounded quartz, fawn Gravel: fine and coarse with a trace of cobble, subangular to subrounded flint, with subrounded to rounded quartz and quartzite, with some rounded sandstone	3.6	5.9
Woolwich and Reading Beds	Clay, very sandy at base, brown mottled blue, stiff, waxy	2.7	8.6
	Sand, fine and medium, silty, clayey, dark red, soft	1.0	9.6
	Clay, very sandy, red, mottled orange, green and dark grey, with pebble beds of black and brown well rounded flints up to 3 cm in diameter, firm	0.9	10.5
Thanet Beds (Thanet Sand)	Sand, fine, clayey, silty, olive green streaked red at top, finely laminated, firm	0.5+	11.0

Grading

Mean for deposit *percentages*			Depth below surface (m)	*percentages*						
Fines	Sand	Gravel		Fines	Sand			Gravel		
				$-\frac{1}{16}$	$+\frac{1}{16}-\frac{1}{4}$	$+\frac{1}{4}-1$	+1–4	+4–16	+16–64	+64
18	80	2	2.3–3.3	17	73	10	–	–	–	–
			3.3–4.3	15	82	3	–	–	–	–
			4.3–4.8	13	80	7	–	–	–	–
			4.8–5.8*	25	70	3	1	1	–	–
			5.8–5.9*	5	10	12	18	22	25	8
			Mean	18	74	5	1	1	1	8

Composition

Depth below surface (m)	Percentages by weight in +4–16 mm fraction					
	Chalk	Flint	Quartz and Quartzite	Fossil debris	Sandstone	Others
2.3–3.3	No pebbles, sand					
3.3–4.3	No pebbles, sand					
4.3–4.8	No pebbles, sand					
4.8–5.8	–	30	30	–	26	14
5.8–5.9	–	61	20	–	12	7
Mean	–	33	29	–	25	13

TL 41 NE 32 **4848 1760** **Spellbrook Farm, Sawbridgeworth** **Block D**

Surface level +58.1 m
Water struck at +41.2 m
Shell (modified) 152 mm diameter
March 1976

Waste 16.0 m
Bedrock 1.0 m+

Log

Geological classification	*Lithology*	*Thickness* m	*Depth* m
	Soil	0.3	0.3
Boulder Clay	Clay, brown, with angular flints and some well rounded quartz, firm	0.7	1.0
	Clay, silty, flinty, chalky, yellow brown, very sandy in parts, soft	1.0	2.0
	Clay, chalky, flinty, silty, brown mottled grey with depth, firm	1.7	3.7
	Clay, chalky, silty, blue grey, stiff	12.3	16.0
Upper Chalk	Chalk, pugged up, white, with blue and black angular flints	1.0+	17.0

TL 41 NE 33 **4856 1676** **½ km West of Tednambury, Sawbridgeworth** **Block D**

Surface level +59.6 m
Water struck at +42.5 m
Shell (modified) 152 mm diameter
March 1976

Waste 24.8 m+

Log

Geological classification	*Lithology*	*Thickness* m	*Depth* m
	Soil	0.2	0.2
Boulder Clay	Clay, chalky, silty, mottled blue and orange, soft	0.7	0.9
	Clay, chalky, silty, brown mottled grey, soft	3.3	4.2
	Clay, silty, chalky, flinty, grey blue, stiff	5.4	9.6
Lacustrine **Deposits**	Clays and silts, laminated, light grey, becomes very silty and sandy from 22.0 m. Noticeably chalky from 23.2 m. Laminae range from 0.1 mm to 1 mm with coarser banding of clay and silt up to 25 mm	15.2+	24.8

TL 41 NE 34 **4967 1915** **Twyford Lock, Thorley** **Block D**

Surface level +67.1 m
Water struck at +57.6 m, +46.6 m and +46.3 m
Shell (modified) 152 mm diameter
June 1976

Overburden 3.9 m
Mineral 9.4 m
Waste 7.7 m+

Log

Geological classification	*Lithology*	*Thickness* m	*Depth* m
	Soil	0.3	0.3
Head	Clay, pebbly, sandy, brown, soft	0.5	0.8
	Clay, very sandy, very pebbly, orange brown, firm	0.8	1.6
	Clay, sandy, silty, chalky, mottled red and brown, and grey blue, stiff	0.4	2.0
	Clay, very sandy, pebbly, orange brown, firm	0.8	2.8
	Clay, silty, mottled brown and grey, with carbonaceous debris, very stiff	1.1	3.9
Glacial Sand and Gravel	Sandy gravel, becomes more sandy and chalky with depth Sand: medium with fine and coarse, predominantly rounded quartz, with angular medium and coarse flint, brown Gravel: fine with coarse, subangular to subrounded flint, with some rounded chalk, and a trace of fossil debris, rounded quartz, quartzite and sandstone	9.4	13.3
Boulder Clay	Clay, chalky, silty, brown, stiff	0.4	13.7
	Clay, chalky, silty, dark blue grey, with some flint pebbles, stiff	0.5	14.2
	Clay, very chalky and silty, light grey, with silt bands throughout and medium to fine chalk sand bands in last metre	6.8+	21.0

Grading

Mean for deposit *percentages*			Depth below surface (m)	*percentages*					
Fines	Sand	Gravel		Fines	Sand			Gravel	
				$-\frac{1}{16}$	$+\frac{1}{16}-\frac{1}{4}$	$+\frac{1}{4}-1$	+1–4	+4–16	+16
9	57	34	3.9–5.2	10	19	33	9	19	10
			5.2–6.2	13	20	28	8	16	15
			6.2–7.5	9	20	24	9	17	21
			7.5–8.5	9	7	23	10	20	31
			8.5–9.6	11	6	26	14	28	15
			9.6–11.0*	6	8	39	18	21	8
			11.0–12.0*	3	6	35	18	31	7
			12.0–13.3*	10	11	44	13	13	9
			Mean	9	12	32	13	20	14

Composition

Depth below surface (m)	Percentages by weight in +4–16 mm fraction					
	Chalk	Flint	Quartz and Quartzite	Fossil debris	Sandstone	Others
3.9–5.2	–	82	7	–	7	4
5.2–6.2	–	91	3	–	2	4
6.2–7.5	–	87	4	1	2	6
7.5–8.5	–	82	7	2	1	8
8.5–9.6	14	74	1	5	4	2
9.6–11.0	12	86	$\frac{1}{2}$	–	$\frac{1}{2}$	1
11.0–12.0	14	73	3	1	5	4
12.0–13.3	10	78	3	2	2	5
Mean	6	83	3	1	3	4

TL 41 NE 35 **4920 1867** **Thorley Street, Thorley** **Block E**

Surface level +52.5 m
Water struck at +50.4 m
Shell (modified) 152 mm diameter
March 1976

Overburden 2.1 m
Mineral 8.2 m
Waste 2.9 m
Bedrock 0.8 m+

Log

Geological classification	*Lithology*	*Thickness* m	*Depth* m
	Soil	0.1	0.1
Alluvium	Clay, sandy, silty, brown, soft, sticky	0.2	0.3
	Peat, very clayey, silty, black, with some fine and medium sand, becomes less clayey with depth	1.8	2.1
Glacial Sand and Gravel (Buried Channel Deposits)	'Clayey' sandy gravel, gravel at top becomes very clayey pebbly sand from 6.0 m. Pebbly, silty, sandy, brown, soft clay bands between 3.2 and 3.7 m and 4.1–4.5 m, and a band of reconstituted chalk paste between 5.8 and 6.0 m Sand: coarse and medium with a trace of fine at top, fine with medium and some coarse at base, predominantly quartz, with some angular coarse and medium flint and rounded chalk Gravel: fine and coarse, finer with depth, angular to subrounded flint, with rounded chalk, and some rounded quartz, quartzite and sandstone, and a trace of fossil debris	8.2	10.3
Boulder Clay	Clay, silty, pebbly, laminated, brown and black, soft	0.1	10.4
	Clay, silty, blue grey, soft	0.4	10.8
	Clay, chalky, grey, soft	2.4	13.2
Upper Chalk	Chalk, angular fragments of chalk and flint in chalk slurry	0.8+	14.0

Grading

Mean for deposit *percentages*			Depth below surface (m)	*percentages*					
Fines	Sand	Gravel		Fines	Sand			Gravel	
				$-\frac{1}{16}$	$+\frac{1}{16}-\frac{1}{4}$	$+\frac{1}{4}-1$	+1–4	+4–16	+16
15	58	27	2.1–3.2*	3	2	12	14	31	38
			3.7–4.1*	2	2	10	17	31	38
			4.5–4.8*	4	2	10	18	40	26
			4.9–5.8*	5	4	14	27	43	7
			6.0–9.0*	26	61	12	1	–	–
			9.0–10.3*	13	35	29	8	10	5
			Mean	15	34	15	9	16	11

Composition

Depth below surface (m)	Percentages by weight in + 4–16 mm fraction					
	Chalk	Flint	Quartz and Quartzite	Fossil debris	Sandstone	Others
2.1–3.2	7	76	9	–	5	3
3.7–4.1	11	74	5	½	6	3½
4.5–4.8	21	50	10	3	7	9
4.9–5.8	32	43	9	2	11	3
6.0–9.0	No pebbles, sand					
9.0–10.3	40	41	6	4	3	6
Mean	25	55	8	2	6	4

TL 41 NE 36 4977 1813 George P.H., Little Hallingbury **Block D**

Surface level +67.1 m
Water struck at +56.1 m
Shell (modified) 152 mm diameter
May 1976

Overburden 1.0 m
Mineral 6.3 m
Waste 8.0 m
Bedrock 1.7 m+

Log

Geological classification	*Lithology*	*Thickness* m	*Depth* m
	Soil	0.4	0.4
	Subsoil, brown pebbly clay with rootlets	0.6	1.0
Head Gravel	'Very clayey' sandy gravel, very sandy between 2.9 m and 4.5 m, alternating bands of hoggin, sands, and silts from 4.5 m to 5.5 m Sand: medium and fine with coarse, predominantly rounded quartz, orange brown Gravel: fine with coarse, angular to subrounded flint, with some quartz, quartzite and sandstone, and a trace of chalk in lower 2.8 m	6.3	7.3
Boulder Clay	Clay, sandy, pebbly, chalky, brown becoming blue grey, firm	3.7	11.0
	Sand, very clayey, silty, grey, soft	0.2	11.2
	Clay, chalky, silty, flinty, brown becoming dark blue grey, stiff	4.1	15.3
Woolwich and Reading Beds	Clay, dark blue black mottled red brown and light blue, becomes brown mottled red and light blue, stiff, waxy	1.5	16.8
	Clay, brown mottled red and light blue, with race nodules, stiff, waxy	0.2+	17.0

Grading

Mean for deposit *percentages*			Depth below surface (m)	*percentages*					
Fines	Sand	Gravel		Fines	Sand			Gravel	
				$-\frac{1}{16}$	$+\frac{1}{16}-\frac{1}{4}$	$+\frac{1}{4}-1$	+1–4	+4–16	+16
26	44	30	1.0–2.0	19	5	19	11	21	25
			2.0–2.9	17	6	17	7	19	34
			2.9–4.0	20	29	36	6	8	1
			4.0–4.5	36	44	15	2	2	1
			4.5–5.5	53	18	11	5	9	4
			5.5–6.5	16	9	16	16	30	13
			6.5–7.3	22	13	18	10	27	10
			Mean	26	16	19	9	17	13

Composition

Depth below surface (m)	Percentages by weight in + 4–16 mm fraction					
	Chalk	Flint	Quartz and Quartzite	Fossil debris	Sandstone	Others
1.0–2.0	–	82	9	–	7	2
2.0–2.9	–	77	13	–	8	2
2.9–4.0	–	85	5	–	7	3
4.0–4.5	–	89	8	–	3	–
4.5–5.5	2	86	6	–	2	4
5.5–6.5	2	79	7	–	7	5
6.5–7.3	7	70	7	–	8	8
Mean	1	81	8	–	6	4

TL 41 NE 37 4939 1732 Millhide Common, Little Hallingbury **Block D**

Surface level +54.3 m
Water struck at +50.0 m
Shell (modified) 152 mm diameter
June 1976

Waste 2.6 m
Bedrock 5.9 m+

Log

Geological classification	*Lithology*	*Thickness* m	*Depth* m
Made Ground	Clay, pebbly, friable, brown, with brick rubble	1.1	1.1
Glacial Sand and Gravel	'Very clayey' sandy gravel, medium sand, with fine and coarse flint gravel	0.7	1.8
Boulder Clay	Clay, silty, chalky, flinty, brown, soft	0.8	2.6
Woolwich and Reading Beds	Clay, brown mottled green and grey, stiff, waxy	1.7	4.3
	Sand, fine, silty, yellow brown, with some white, brown and black well rounded flints, firm	0.6	4.9
Thanet Beds (Thanet Sand)	Sand, fine, silty, grey with a pink tinge mottled red, brown and green, with hard pyritic sand nodules, firm	0.5	5.4
	Sand, fine, silty, grey green mottled red and rarely brown, soft	3.1+	8.5

TL 41 NE 38 4943 1624 Oaks and Bushes, Little Hallingbury **Block D**

Surface level +54.0 m
Water struck at +48.4 m and +45.5 m
Shell (modified) 152 mm diameter
June 1976

Overburden 0.8 m
Mineral 5.7 m
Bedrock 2.2 m+

Log

Geological classification	*Lithology*	*Thickness* m	*Depth* m
	Soil	0.3	0.3
Head	Clay, pebbly becomes very pebbly, orange brown, firm	0.5	0.8
Glacial Sand and Gravel	'Clayey' gravel Sand: medium with some coarse and fine, predominantly rounded quartz Gravel: fine with coarse, subangular to rounded flint, with some rounded quartz, quartzite and sandstone	5.7	6.5
London Clay (Basement Bed)	Clay, silty, fine sandy, laminated, brown, firm	0.3	6.8
	Silt, clayey, fine sandy, drab olive green, with wood fragments and glauconite, well rounded black fine gravel flints in lower 0.1 m	0.3	7.1
Woolwich and Reading Beds	Clay, very silty, laminated with fine sand, pale blue green mottled green and brown, stiff	0.7	7.8
	Sand fine, with coarse silt, clayey, yellow brown mottled pale green and light brown, stiff	0.9+	8.7

Grading

Mean for deposit *percentages*			Depth below surface (m)	*percentages*					
Fines	Sand	Gravel		Fines	Sand			Gravel	
				$-\frac{1}{16}$	$+\frac{1}{16}-\frac{1}{4}$	$+\frac{1}{4}-1$	+1–4	+4–16	+16
10	43	47	0.8–1.8	8	5	14	10	27	36
			1.8–2.6	11	8	26	11	28	16
			2.6–3.8	27	16	23	6	13	15
			3.8–4.8	5	7	22	14	35	17
			4.8–5.6	3	6	29	10	35	17
			5.6–6.5*	3	11	32	9	28	17
			Mean	10	9	24	10	27	20

Composition

Depth below surface (m)	Percentages by weight in + 4–16 mm fraction					
	Chalk	Flint	Quartz and Quartzite	Fossil debris	Sandstone	Others
0.8–1.8	–	71	20	–	4	5
1.8–2.6	Sample missing					
2.6–3.8	–	63	22	–	11	4
3.8–4.8	–	67	21	–	5	7
4.8–5.6	–	74	15	–	6	5
5.6–6.5	–	77	11	–	9	3
Mean	–	70	18	–	7	5

TL 41 NE 39 4917 1505 Railway Inn, Sawbridgeworth **Block E**

Surface level +49.0 m
Water struck at +47.0 m and +39.5 m
Shell (modified) 152 mm diameter
May 1976

Waste 2.3 m
Bedrock 7.8 m+

Log

Geological classification	*Lithology*	*Thickness* m	*Depth* m
	Made ground	0.3	0.3
Alluvium	Clay, very sandy, pebbly, silty, brown black, soft	1.4	1.7
Glacial Sand and Gravel	Sand, pebbly, brown and black flints in a sandy, very clayey matrix	0.6	2.3
London Clay	Silt, clayey, fine sandy, dark blue grey, some green mottling with depth; more clayey with depth, band of shell debris and mudstone at 3.7 m	1.6	3.9
	Clay, silty, fine sandy, laminated, grey green, shell debris, wood fragments and claystones at base	1.9	5.8
London Clay (Basement Bed)	Clay, laminated, grey green, bioturbated	3.1	8.9
Woolwich and Reading Beds	Clay, dark blue grey, with many well preserved cemented bivalve shells, small well rounded flints, glauconitic shell band at top	0.6	9.5
	Clay, mottled red, brown, blue, green and white, stiff, waxy	0.6+	10.1

TL 41 SW 12 4077 1450 Near Filletts Farm, Hunsdon **Block B**

Surface level +72.2 m
Water struck at +69.5 m
Shell and Auger 152 mm diameter
October 1975

Overburden 3.4 m
Mineral 1.2 m
Waste 4.9 m
Mineral 1.8 m
Bedrock 1.7 m+

Log

Geological classification	*Lithology*	*Thickness* m	*Depth* m
	Soil	0.4	0.4
Boulder Clay	Clay, pebbly, yellow brown, soft	0.7	1.1
	Clay, silty, pebbly, chalky, mottled yellow brown and light blue, soft	1.6	2.7
	Silt, clayey, sandy, yellow, laminated	0.7	3.4
Glacial Sand and Gravel	**a** 'Clayey' gravel Sand: medium with fine and coarse, predominantly rounded quartz, with coarse angular flints, yellow brown Gravel: coarse and fine with cobbles at base, angular to rounded flint, with rounded quartz and quartzite, and a trace of rounded sandstone and chalk	1.2	4.6
Boulder Clay	Clay, silty, chalky, with some flint pebbles, blue grey, stiff	4.7	9.3
	Clay, silty, sandy, brown, very pebbly at base	0.2	9.5
Glacial Sand and Gravel	**b** 'Very clayey' gravel, clay bands throughout Sand: coarse, medium and fine quartz, with some angular coarse flint, yellow brown Gravel: coarse and fine, angular to subrounded flint, with rounded quartz and quartzite, and a trace of sandstone	1.8	11.3
London Clay	Clay, sandy, silty, yellow brown, soft, laminated	0.3	11.6
	Clay, silty, fine, sandy, olive grey, soft	0.4	12.0
	Clay, silty, olive grey, firm	1.0+	13.0

Grading

	Mean for deposit *percentages*			Depth below surface (m)	*percentages*						
	Fines	Sand	Gravel		Fines	Sand			Gravel		
					$-\frac{1}{16}$	$+\frac{1}{16}-\frac{1}{4}$	$+\frac{1}{4}-1$	+1–4	+4–16	+16–64	+64
a	11	21	68	3.4–4.4*	13	4	11	6	31	35	–
				4.4–4.6*	5	4	12	7	29	39	4
				Mean	11	4	11	6	31	36	1
b	20	32	48	9.5–10.5	14	7	11	11	26	31	–
				10.5–11.3	27	11	14	11	22	15	–
				Mean	20	9	12	11	24	24	–

Composition

	Depth below surface (m)	Percentages by weight in +4–16 mm fraction					
		Chalk	Flint	Quartz and Quartzite	Fossil debris	Sandstone	Others
a	3.4–4.4	2	68	24	–	4	2
	4.4–4.6	–	69	25	–	3	3
	Mean	2	68	24	–	4	2
b	9.5–10.5	–	73	20	–	7	–
	10.5–11.3	–	72	19	–	8	1
	Mean	–	73	20	–	7	–

TL 41 SW 13 4076 1372 Filletts Farm, Hunsdon **Block B**

Surface level +69.0 m
Water not struck
Shell (modified) 152 mm diameter
July 1972

Overburden 3.5 m
Mineral 1.8 m
Waste 11.8 m
Mineral 8.6 m
Bedrock 0.4 m+

Log

Geological classification	*Lithology*	*Thickness* m	*Depth* m
	Soil	0.4	0.4
Boulder Clay	Clay, sandy, flinty at base, brown, soft	3.1	3.5
Glacial Sand and Gravel	**a** Gravel Sand: medium and coarse with a trace of fine, predominantly quartz, with some angular coarse flint, yellow brown Gravel: coarse with fine, subrounded to rounded flint, with some rounded quartz and quartzite	1.8	5.3
Boulder Clay	Clay, silty, chalky, brown, soft	2.3	7.6
	Clay, very sandy, very pebbly, brown, firm	0.7	8.3
	Clay, silty, chalky, grey becoming brown, firm becoming soft	3.9	12.2
Glacial Sand and Gravel	**b** 'Very clayey' pebbly sand Sand: medium with a trace of fine and coarse, predominantly rounded quartz, orange Gravel: coarse and fine, well rounded to rounded flint	2.2	14.4
Boulder Clay	Clay, silty, chalky, with bands of fine chalky sand, grey, soft	1.2	15.6
	Clay, chalky, brown, soft	1.5	17.1
Glacial Sand and Gravel	**c** 'Very clayey' gravel, pebbly, very clayey medium and fine sand between 20.1 and 22.1 m Sand: medium with coarse and some fine, predominantly quartz, with angular coarse flint, brown Gravel: fine with coarse, angular to well rounded flint, with some subrounded quartz and quartzite	8.6	25.7
Woolwich and Reading Beds	Sand, fine, silty, clayey, mottled red and grey, firm	0.4+	26.1

Grading

	Mean for deposit *percentages*			Depth below surface (m)	*percentages*					
	Fines	Sand	Gravel		Fines	Sand			Gravel	
					$-\frac{1}{16}$	$+\frac{1}{16}-\frac{1}{4}$	$+\frac{1}{4}-1$	$+1-4$	$+4-16$	$+16$
a	6	26	68	3.5–4.5	Sample missing					
				4.5–5.3	6	2	13	11	28	40
				Mean	6	2	13	11	28	40
b	30	59	11	12.2–13.2	41	9	44	1	2	3
				13.2–14.4	21	6	51	6	7	9
				Mean	30	7	48	4	5	6
c	21	36	43	17.1–18.1	15	4	20	11	25	25
				18.1–19.1	Sample missing					
				19.1–20.1	21	3	17	13	28	18
				20.1–21.1	15	5	22	12	33	13
				21.1–22.1	39	23	33	3	2	–
				22.1–23.1	Sample missing					
				23.1–24.1	23	–	9	10	28	30
				24.1–25.1	12	2	14	14	34	24
				25.1–25.7	Not sampled					
				Mean	21	6	19	11	25	18

TL 41 SW 14 4037 1285 Halfway House, Stanstead Abbots **Block B**

Surface level +69.8 m
Water struck at +63.0 m and +55.0 m
Shell (modified) 152 mm diameter
June 1972

Overburden 4.8 m
Mineral 5.1 m
Waste 4.9 m
Mineral 9.5 m
Waste 2.8 m
Bedrock 1.4 m+

Log

Geological classification	*Lithology*	*Thickness* m	*Depth* m
	Soil	0.1	0.1
Boulder Clay	Clay, chalky, light brown, soft	0.4	0.5
	Clay, chalky, grey, soft	2.9	3.4
	Clay, chalky, brown, soft	1.4	4.8
Glacial Sand and Gravel	**a** 'Clayey' gravel Sand: medium with some coarse and a trace of fine, predominantly quartz, with some angular coarse flint, yellow brown Gravel: fine with some coarse, subangular to well rounded flint, with rounded quartz and quartzite	5.1	9.9
Boulder Clay	Clay, sandy, chalky, brown becoming grey, firm	4.9	14.8
Glacial Sand and Gravel	**b** 'Clayey gravel' Sand: coarse with medium and a trace of fine, predominantly angular coarse flint, with rounded quartz, yellow Gravel: coarse with some fine, angular to subrounded flint, with some quartz, quartzite and sandstone, and a trace of chalk and fossil debris	9.5	24.3
Boulder Clay	Clay, chalky, grey, soft	2.8	27.1
Woolwich and Reading Beds	Clay, sandy, mottled red and green, firm	1.0	28.1
	Sand, medium, clayey, green, soft	0.4+	28.5

Grading

	Mean for deposit *percentages*			Depth below surface (m)	*percentages*					
	Fines	Sand	Gravel		Fines	Sand			Gravel	
					$-\frac{1}{16}$	$+\frac{1}{16}-\frac{1}{4}$	$+\frac{1}{4}-1$	+1–4	+4–16	+16
a	13	35	52	4.8–5.8	17	2	17	14	29	21
				5.8–6.8	10	3	16	14	38	19
				6.8–7.8*	10	3	28	9	30	20
				7.8–8.8*	Sample missing					
				8.8–9.9*	Sample missing					
				Mean	13	3	20	12	32	20
b	14	41	45	14.8–15.8*	18	2	17	20	22	21
				15.8–16.8*	7	2	15	20	16	40
				16.8–18.8*	Sample missing					
				18.8–20.8*	16	2	13	28	13	28
				20.8–21.8*	25	2	16	29	12	16
				21.8–22.8*	Sample missing					
				22.8–23.8*	8	2	15	22	19	34
				23.8–24.3*	7	4	19	19	17	34
				Mean	14	2	16	23	17	28

TL 41 SW 15 **4067 1226** **Olives Farm, Hunsdon** **Block B**

Surface level +68.0 m
Water struck at +60.4 m
Shell (modified) 152 mm diameter
March 1976

Waste 18.0 m
Bedrock 1.0 m+

Log

Geological classification	*Lithology*	*Thickness* m	*Depth* m
	Soil	0.3	0.3
Boulder Clay	Clay, silty, flinty, brown, soft	0.2	0.5
	Clay, silty, chalky, brown becoming dark brown, stiff	4.3	4.8
	Clay, silty, chalky, blue grey, stiff	2.8	7.6
	'Clayey' sand, silty, chalky, pale grey, soft	0.1	7.7
	Clay, chalky, dark blue grey, very hard, shaley	2.1	9.8
	Clay, chalky, dark blue grey becoming dark brown, stiff	3.8	13.6
	Clay, silty, chalky, flinty, brown, soft	0.4	14.0
Glacial Sand and Gravel	'Clayey' gravel, sandy at top becoming coarser with depth Sand: medium with coarse and fine, predominantly rounded quartz, with some chalk, pale yellow Gravel: fine with coarse, rounded to angular flint, with some rounded quartz and quartzite, and a trace of rounded chalk and sandstone	4.0	18.0
London Clay	Clay, silty, sandy, mottled brown, orange and grey, firm	0.4	18.4
	Clay, silty, fine sandy, drab olive green, with pyrite nodules and decalcified shell debris, firm	0.6+	19.0

Grading

Mean for deposit *percentages*			Depth below surface (m)	*percentages*					
Fines	Sand	Gravel		Fines	Sand			Gravel	
				$-\frac{1}{16}$	$+\frac{1}{16}-\frac{1}{4}$	$+\frac{1}{4}-1$	+1–4	+4–16	+16
11	42	47	14.0–15.0	11	23	22	11	22	11
			15.0–16.0	12	8	19	16	29	16
			16.0–17.0	4	5	17	17	32	25
			17.0–18.0	18	5	12	14	31	20
			Mean	11	10	18	14	29	18

Composition

Depth below surface (m)	Percentages by weight in + 4–16 mm fraction					
	Chalk	Flint	Quartz and Quartzite	Fossil debris	Sandstone	Others
14.0–15.0	5	70	14	–	6	5
15.0–16.0	3	57	16	–	16	8
16.0–17.0	3	69	14	–	9	5
17.0–18.0	2	75	13	–	5	5
Mean	3	68	14	–	9	6

TL 41 SW 16 4048 1151 ½ km West of Briggens Farm, Stanstead Abbots **Block B**

Surface level +59.9 m
Water not struck
Shell (modified) 152 mm diameter
October 1972

Overburden 8.3 m
Mineral 3.7 m
Bedrock 3.0 m+

Log

Geological classification	*Lithology*	*Thickness* m	*Depth* m
	Soil and subsoil	0.6	0.6
Boulder Clay	Clay, chalky, brown, sand band at 1.8 m, soft	1.4	2.0
	Clay, chalky, grey, soft	5.8	7.8
	Clay, chalky, brown, soft	0.5	8.3
Glacial Sand and Gravel	Sandy gravel Sand: medium with a trace of coarse and fine, predominantly quartz, brown Gravel: fine and coarse, subangular to rounded flint, with some quartz and quartzite	3.7	12.0
London Clay	Clay, sandy, silty, brown, stiff	0.6	12.6
	Clay, silty, grey to blue grey, stiff	2.4+	15.0

Grading

Mean for deposit *percentages*			Depth below surface (m)	*percentages*					
Fines	Sand	Gravel		Fines	Sand			Gravel	
				$-\frac{1}{16}$	$+\frac{1}{16}-\frac{1}{4}$	$+\frac{1}{4}-1$	+1–4	+4–16	+16
6	59	35	8.3–9.3	Sample missing					
			9.3–10.3	8	13	61	4	8	6
			10.3–11.3	Sample missing					
			11.3–12.0	3	4	13	16	31	33
			Mean	6	9	41	9	18	17

TL 41 SW 17 **4072 1065** **Gatehouse Lodge, Briggens Park, Stanstead Abbots** **Block E**

Surface level +31.3 m
Water struck at +27.1 m
Shell 152 mm diameter
September 1975

Overburden 2.2 m
Mineral 5.9 m
Bedrock 1.4 m+

Log

Geological classification	*Lithology*	*Thickness* m	*Depth* m
	Soil, black, silty	0.2	0.2
Alluvium	Clay, silty, sandy, grey brown, very soft	1.4	1.6
	Peat, clayey to very clayey, silty, fine sandy, black, many wood fragments, soft	0.6	2.2
	Gravel, becomes coarser with depth Sand: medium and coarse with some fine, quartz with flint, yellow Gravel: fine and coarse, angular to rounded flint, with quartz, and a trace of sandstone, chalk, and ironstone	5.9	8.1
Thanet Beds (Thanet Sand)	'Very clayey' sand, fine with a trace of medium and coarse, drab olive green, glauconitic	0.5	8.6
(Bullhead Bed)	Sand, fine with a trace of medium, with angular black and brown flints in a pale green calcite concretionary cement at base	0.5	9.1
Upper Chalk	Chalk, pugged up, white streaked brown	0.4+	9.5

Grading

Mean for deposit percentages			Depth below surface (m)	percentages					
Fines	Sand	Gravel		Fines	Sand			Gravel	
				$-\frac{1}{16}$	$+\frac{1}{16}-\frac{1}{4}$	$+\frac{1}{4}-1$	+1–4	+4–16	+16
6	35	59	2.2–3.2	16	10	24	9	24	17
			3.2–4.2	15	8	20	16	29	12
			4.2–5.4*	2	4	22	16	34	22
			5.4–6.2*	1	3	14	15	37	30
			6.2–7.2*	1	2	10	11	41	35
			7.2–8.1*	1	2	9	13	41	35
			Mean	6	5	17	13	34	25
			Bedrock analyses						
21	79	—	8.1–8.6*	21	73	3	3	–	–

Composition

Depth below surface (m)	Percentages by weight in + 4–16 mm fraction					
	Chalk	Flint	Quartz and Quartzite	Fossil debris	Sandstone	Others
2.2–3.2	1	81	12	–	2	4
3.2–4.2	3	72	15	–	3	7
4.2–5.4	4	73	12	–	6	5
5.4–6.2	2	75	17	–	3	3
6.2–7.2	2	72	16	–	6	4
7.2–8.1	3	63	25	–	4	5
Mean	3	73	16	–	4	4

TL 41 SW 18 4182 1372 The Rectory, Hunsdon **Block B**

Surface level +73.3 m
Water not struck
Shell 152 mm diameter
October 1975

Overburden 15.0 m
Mineral 5.2 m
Bedrock 1.3 m+

Log

Geological classification	*Lithology*	*Thickness* m	*Depth* m
	Soil	0.3	0.3
Boulder Clay	Clay, chalky, silty, pebbly, brown, soft	3.9	4.2
	'Very clayey' gravel Sand: fine, medium and coarse, predominantly quartz, with some flint, brown Gravel: fine and coarse, angular flint, with rounded quartz, and a trace of sandstone, chalk and fossil debris	1.0	5.2
	Clay, very chalky and flinty, grey blue becoming blue black with depth, stiff	9.8	15.0
Glacial Sand and Gravel	'Clayey' pebbly sand; 'very clayey' at top, becomes sandy gravel at base Sand: medium and fine, with a trace of angular coarse flint, yellow brown mottled grey Gravel: fine and coarse, angular to rounded flint, with a trace of rounded quartz, quartzite and sandstone	5.2	20.2
London Clay	Clay, silty, sandy in parts with black carbonaceous partings, olive grey	1.3+	21.5

Grading

Mean for deposit *percentages*			Depth below surface (m)	*percentages*					
Fines	Sand	Gravel		Fines	Sand			Gravel	
				$-\frac{1}{16}$	$+\frac{1}{16}-\frac{1}{4}$	$+\frac{1}{4}-1$	+1–4	+4–16	+16
26	22	52	4.2–5.2	26	8	7	7	25	27
19	66	15	15.0–16.0	58	25	13	1	1	2
			16.0–17.0	16	17	48	4	9	5
			17.0–18.0	10	18	58	7	5	2
			18.0–19.0	5	38	32	4	12	9
			19.0–20.2	6	44	13	6	14	17
			Mean	19	29	32	5	8	7

Composition

Depth below surface (m)	Percentages by weight in + 4–16 mm fraction					
	Chalk	Flint	Quartz and Quartzite	Fossil debris	Sandstone	Others
4.2–5.2	4	64	25	1	4	3
15.0–16.0	1	86	9	–	3	1
16.0–17.0	–	80	11	–	8	1
17.0–18.0	–	81	9	–	9	1
18.0–19.0	–	77	19	–	3	1
19.0–20.2	–	80	10	–	1	3
Mean	–	81	13	–	5	1

TL 41 SW 19 4172 1289 St. Dunstons Church, Hunsdon **Block B**

Surface level +67.6 m
Water struck at +64.0 m and 61.3 m
Auger and extending Kelly 203 mm diameter
December 1975

Overburden 2.5 m
Mineral 2.0 m
Waste 8.7 m
Mineral 2.9 m+

Log

Geological classification	*Lithology*	*Thickness* m	*Depth* m
	Soil	0.2	0.2
Boulder Clay	Clay, chalky, pebbly, yellow brown, soft	1.6	1.8
	Clay, chalky, light grey, soft	0.7	2.5
Glacial Sand and Gravel	**a** 'Very clayey' gravel Sand: medium, fine and coarse, predominantly quartz, with coarse angular flint, brown Gravel: fine and coarse, rounded with some angular flint and rounded chalk, with some rounded quartz and sandstone, and a trace of fossil debris	2.0	4.5
Boulder Clay	Clay, silty, sandy, chalky, brown, soft	0.9	5.4
	Clay, chalky, dark blue, stiff	0.9	6.3
	Clay, very chalky, silty and sandy, light blue, very soft	1.3	7.6
	Clay, silty, chalky, blue grey, firm to stiff	5.1	12.7
	Clay, chalky, sandy, pebbly, brown, soft	0.5	13.2
Glacial Sand and Gravel	**b** 'Clayey' pebbly sand, becoming more pebbly with depth Sand: fine and medium with a trace of coarse, yellow brown Gravel: fine and coarse, angular flint, with rounded chalk and quartz, and a trace of sandstone and fossil debris	2.9+	16.1

Grading

	Mean for deposit *percentages*			Depth below surface (m)	*percentages*					
	Fines	Sand	Gravel		Fines	Sand			Gravel	
					$-\frac{1}{16}$	$+\frac{1}{16}-\frac{1}{4}$	$+\frac{1}{4}-1$	+1–4	+4–16	–16
a	23	29	48	2.5–3.5	21	9	11	8	27	24
				3.5–4.5*	25	9	11	9	25	21
				Mean	23	9	11	9	26	22
b	15	78	7	13.2–14.9	15	46	35	2	1	1
				14.9–16.1	14	32	35	4	8	7
				Mean	15	40	35	3	4	3

Composition

	Depth below surface (m)	Percentages by weight in +4–16 mm fraction					
		Chalk	Flint	Quartz and Quartzite	Fossil debris	Sandstone	Others
a	2.5–3.5	19	55	12	2	7	5
	3.5–4.5	34	47	6	1	7	5
	Mean	26½	51	9	1½	7	5
b	13.2–14.9	Small sample – all flint					
	14.9–16.1	10	65	12	1	6	6

TL 41 SW 20 4172 1289 St. Dunstons Church, Hunsdon **Block B**

Surface level +67.6 m
Water struck at +63.8 m and +61.3 m
Shell 152 mm diameter
December 1975

Overburden 2.4 m
Mineral 1.5 m
Waste 8.1 m
Mineral 4.4 m
Waste 1.4 m
Mineral 3.2 m+

Log

Geological classification	*Lithology*	*Thickness* m	*Depth* m
	Soil	0.4	0.4
Boulder Clay	Clay, chalky, pebbly, yellow brown, soft	1.4	1.8
	Clay, chalky, very sandy, brown, soft	0.6	2.4
Glacial Sand and Gravel	**a** 'Very clayey' gravel Sand: medium and coarse with fine, predominantly quartz, with coarse angular flint, brown Gravel: coarse and fine, well rounded to angular flint, with rounded chalk and quartz, with a trace of sandstone and fossil debris	1.5	3.9
Boulder Clay	Clay, sandy, silty, chalky, brown, soft	1.0	4.9
	Clay, silty, chalky, grey, firm to stiff	1.4	6.3
	Clay, very silty and chalky, pebbly, grey, very soft	1.3	7.6
	Clay, chalky, silty, blue grey, stiff	1.9	9.5
	Clay, sandy, silty, brown, soft	0.3	9.8
	Clay, chalky, silty, blue to blue grey, stiff	1.8	11.6
	Clay, sandy, silty, pebbly, brown, firm	0.4	12.0
Glacial Sand and Gravel	**b** 'Clayey' sand Sand: fine with medium and a trace of coarse, yellow Gravel: fine and coarse, angular flint and rounded chalk, with a trace of quartz and sandstone	4.4	16.4
Boulder Clay	Clay, chalky, pebbly, sandy, silty, brown, firm	0.6	17.0
	Clay, pebbly, chalky, dark blue, stiff	0.5	17.5
	Clay, very sandy, pebbly, brown, firm	0.3	17.8
Glacial Sand and Gravel	**c** Pebbly sand, more pebbly with depth Sand: medium with some fine and coarse, yellow brown Gravel: fine and coarse, angular to subangular flint, with rounded quartz, and a trace of sandstone and chalk	3.2+	21.0

Grading

	Mean for deposit *percentages*			Depth below surface (m)	*percentages*					
	Fines	Sand	Gravel		Fines	Sand			Gravel	
					$-\frac{1}{16}$	$+\frac{1}{16}-\frac{1}{4}$	$+\frac{1}{4}-1$	+ 1–4	+ 4–16	+ 16
a	22	32	46	2.4–3.9	22	7	15	10	22	24
b	10	84	4	12.0–13.0	12	76	11	1	–	–
				13.0–14.0	12	57	28	2	1	–
				14.0–14.7	11	44	39	2	3	1
				14.7–15.7	7	40	34	5	7	7
				15.7–16.4	8	50	39	2	1	–
				Mean	10	54	29	3	2	2
c	4	72	24	17.8–18.8	6	17	65	5	6	1
				18.8–19.8	3	6	51	7	16	17
				19.8–20.8	2	10	50	7	17	14
				20.8–21.0	Not sampled					
				Mean	4	11	55	6	13	11

Composition

	Depth below surface (m)	Percentages by weight in + 4–16 mm fraction					
		Chalk	Flint	Quartz and Quartzite	Fossil debris	Sandstone	Others
a	2.4–3.9	25	51	11	2	6	5
b	12.0–13.0	Small sample – all flint					
	13.0–14.0	13	67	10	–	–	10
	14.0–14.7	10	58	13	3	9	7
	14.7–15.7	11	64	11	1	7	6
	15.7–16.4	Small sample – all flint					
	Mean	11	64	11	1	5	8
c	17.8–18.8	2	84	7	1	2	3
	18.8–19.8	2	79	13	–	2	4
	19.8–20.8	–	71	17	–	5	7
	Mean	1	78	12	1	3	5

TL 41 SW 21 4178 1165 Brickhouse Farm, Hunsdon **Block B**

Surface level +55.5 m
Water not struck
Shell 152 mm diameter
September 1975

Overburden 1.2 m
Mineral 2.0 m
Waste 2.3 m
Mineral 2.9 m
Bedrock 0.3 m+

Log

Geological classification	*Lithology*	*Thickness* m	*Depth* m
	Soil	0.2	0.2
Boulder Clay	Clay, chalky, flinty, brown mottled yellow, soft	1.0	1.2
Glacial Sand and Gravel	**a** Sand Sand: fine and medium with a trace of coarse, yellow brown Gravel: coarse and fine, angular flint, with rounded chalk and quartz, with a trace of sandstone	2.0	3.2
	Silt, chalky, fine sandy, yellow, hard	1.0	4.2
	'Very clayey' sand, trace of pebbles Sand: fine and medium with a trace of coarse, yellow brown Gravel: fine and coarse, angular flint, with a trace of chalk, quartz and sandstone	0.6	4.8
Boulder Clay	Clay, chalky, flinty, olive grey becoming blue grey, firm	0.7	5.5
Glacial Sand and Gravel	**b** 'Clayey' sandy gravel, 'very clayey' pebbly sand becoming 'clayey' gravel Sand: medium, fine and coarse, brown to yellow brown Gravel: fine with coarse, angular flint, with rounded quartz,with a trace of chalk, sandstone and fossil debris	2.9	8.4
London Clay	Clay, silty, becoming sandy, mottled brown, mauve and blue at top, becoming blue grey, stiff	0.3+	8.7

Grading

	Mean for deposit *percentages*			Depth below surface (m)	*percentages*					
	Fines	Sand	Gravel		Fines	Sand			Gravel	
					$-\frac{1}{16}$	$+\frac{1}{16}-\frac{1}{4}$	$+\frac{1}{4}-1$	+1–4	+4–16	+16
a	7	90	3	1.2–2.2	5	68	22	1	–	4
				2.2–3.2	8	27	60	2	2	1
				Mean	7	47	41	2	1	2
	35	62	3	4.2–4.8	35	32	28	2	2	1
b	18	46	36	5.5–6.5	34	29	23	5	6	3
				6.5–7.5	6	12	17	11	38	16
				7.5–8.4	16	8	18	18	24	26
				Mean	18	16	19	11	22	14

Composition

	Depth below surface (m)	Percentages by weight in + 4–16 mm fraction					
		Chalk	Flint	Quartz and Quartzite	Fossil debris	Sandstone	Others
a	1.2–2.2	29	43	14	–	–	14
	2.2–3.2	7	66	13	–	10	4
	Mean	18	54½	13½	–	5	9
	4.2–4.8	13	61	13	–	8	5
b	5.5–6.5	1	66	21	2	4	6
	6.5–7.5	Sample missing					
	7.5–8.4	½	67	21½	–	5	6
	Mean	1	67	21	1	4	6

TL 41 SW 22 4135 1062 Hunsdon Mead, Briggens Park, Roydon **Block E**

Surface level +33.0 m
Water struck at +31.5 m and +30.3 m
Shell 152 mm diameter
September 1975

Overburden 1.7 m
Mineral 4.8 m
Bedrock 7.0 m+

Log

Geological classification	*Lithology*	*Thickness* m	*Depth* m
	Soil	0.2	0.2
Alluvium	Clay, silty, sandy, brown, with flint pebbles, soft	1.5	1.7
	a Gravel, clayey and peaty at top Sand: medium and coarse with some fine, chalky, brown Gravel: fine and coarse, angular flint, with rounded quartz and quartzite, and a trace of rounded chalk and sandstone	4.8	6.5
Thanet Beds (Thanet Sand)	**b** 'Very clayey' sand, a trace of flint pebbles at top Sand: fine with a trace of medium and coarse, glauconitic, drab olive green mottled grey, brown and red in parts	6.3	12.8
(Bullhead Bed)	Sand, silty, predominantly fine with some medium, clayey with depth, many flint pebbles. Cemented black and brown angular flints in pale green calcitic matrix at base	0.5	13.3
Upper Chalk	Chalk, gritty, white to grey, with black and blue angular flints, soft	0.2+	13.5

Grading

	Mean for deposit *percentages*			Depth below surface (m)	*percentages*					
	Fines	Sand	Gravel		Fines	Sand			Gravel	
					$-\frac{1}{16}$	$+\frac{1}{16}-\frac{1}{4}$	$+\frac{1}{4}-1$	+1–4	+4–16	+16
a	4	29	67	1.7–2.7*	14	4	17	10	29	26
				2.7–3.7*	1	1	8	11	40	39
				3.7–4.7*	1	2	11	14	39	33
				4.7–5.7*	1	1	14	14	43	27
				5.7–6.5*	5	6	21	14	27	27
				Mean	4	3	14	12	36	31
				Bedrock analyses						
b	32	67	1	6.5–7.5*	48	44	4	1	2	1
				7.5–8.5*	32	66	2	–	–	–
				8.5–9.5*	39	55	2	2	2	–
				9.5–10.5*	26	70	2	2	–	–
				10.5–11.5*	27	70	2	1	–	–
				11.5–12.8*	25	73	2	1	–	–
				Mean	32	64	2	1	1	–

Composition

	Depth below surface (m)	Percentages by weight in +4–16 mm fraction					
		Chalk	Flint	Quartz and Quartzite	Fossil debris	Sandstone	Others
a	1.7–2.7	1	73	16	–	9	1
	2.7–3.7	–	76	18	–	4	2
	3.7–4.7	2	80	11	–	4	3
	4.7–5.7	2	81	11	–	3	3
	5.7–6.5	1	79	16	–	3	1
	Mean	1	78	14	–	5	2

TL 41 SW 23 **4262 1449** **Eastwick Woods, Hunsdon** **Block B**

Surface level +78.3 m
Water struck at +73.5 m
Shell 152 mm diameter
October 1975

Waste 18.7 m+

Log

Geological classification	*Lithology*	*Thickness* m	*Depth* m
	Soil	0.4	0.4
Boulder Clay	Clay, silty, chalky, mottled blue-grey and brown, soft	4.4	4.8
Glacial Sand and Gravel	Clay, silty, very sandy and gravelly, fine and coarse chalk and angular black flint gravel, with fine, medium and coarse chalk sand, with a trace of quartz, fossil debris and sandstone	1.0	5.8
Boulder Clay	Silt, laminated with fine chalky sand bands, some chalk pellets, grey to olive grey with green mottling, soft	2.3	8.1
	Clay, silty, chalky, some flint and quartz pebbles, blue grey, stiff	10.6+	18.7

Grading

Mean for deposit *percentages*			Depth below surface (m)	*percentages*					
Fines	Sand	Gravel		Fines	Sand			Gravel	
				$-\frac{1}{16}$	$+\frac{1}{16}-\frac{1}{4}$	$+\frac{1}{4}-1$	+1–4	+4–16	+16
41	27	32	4.8–5.8*	41	7	11	9	17	15

Composition

Depth below surface (m)	Percentages by weight in +4–16 mm fraction					
	Chalk	Flint	Quartz and Quartzite	Fossil debris	Sandstone	Others
4.8–5.8	68	14	10	3	5	–

TL 41 SW 24 **4259 1343** **Eastwick Hall Farm, Eastwick** **Block C**

Surface level +69.9 m
Water struck at +53.7 m
Shell 152 mm diameter
November 1975

Waste 18.3 m
Bedrock 1.1 m+

Log

Geological classification	*Lithology*	*Thickness* m	*Depth* m
	Soil	0.2	0.2
Head	Subsoil	0.4	0.6
	Clay, silty, sandy, very pebbly, yellow, soft	0.4	1.0
	Clay, very sandy, pebbly, yellow brown, soft	0.5	1.5
Boulder Clay	Clay, chalky, silty, yellow brown, soft	0.5	2.0
	Sand, fine, with chalky medium, brown, soft	0.5	2.5
	Clay, chalky, silty, flinty, grey, firm	1.2	3.7
	Clay, pebbly, chalky, silty, sandy, dark grey becoming blue grey and black, stiff	10.7	14.4
	Clay, sandy, chalky, brown, soft	0.2	14.6
Glacial Sand and Gravel	Gravel Sand: coarse and medium with some fine, predominantly quartz, with angular coarse flint Gravel: fine with coarse, angular to subrounded flint, with some rounded quartz, and a trace of rounded sandstone	3.3	17.9
London Clay	Clay, laminated, brown mottled dark grey, soft	0.3	18.6
	Clay, silty, sandy, olive grey, with pyrite nodules and decalcified shell debris, firm	0.8+	19.4

Grading

Mean for deposit *percentages*			Depth below surface (m)	*percentages*					
Fines	Sand	Gravel		Fines	Sand			Gravel	
				$-\frac{1}{16}$	$+\frac{1}{16}-\frac{1}{4}$	$+\frac{1}{4}-1$	+1–4	+4–16	+16
2	41	57	14.6–15.6	2	6	16	18	29	29
			15.6–16.6*	2	4	13	23	42	16
			16.6–17.9*	2	4	23	17	31	23
			Mean	2	4	18	19	34	23

Composition

Depth below surface (m)	Percentages by weight in + 4–16 mm fraction					
	Chalk	Flint	Quartz and Quartzite	Fossil debris	Sandstone	Others
14.6–15.6	–	84	10	–	4	2
15.6–16.6	–	71	19	–	8	2
16.6–17.9	–	72	22	–	4	2
Mean	–	75	18	–	5	2

TL 41 SW 25 4272 1236 ½ km N.E. of Brickhouse Farm, Eastwick **Block B**

Surface level +64.9 m
Water not struck
Shell 152 mm diameter
September 1975

Waste 18.6 m+

Log

Geological classification	*Lithology*	*Thickness* m	*Depth* m
	Soil	0.1	0.1
Boulder Clay	Clay, silty, flinty, brown, stiff	0.2	0.3
	Clay, silty, very chalky, some flints, brown becoming yellow, stiff	3.8	4.1
	Clay, silty, very chalky, rare flints, mottled grey and brown, stiff	0.9	5.0
	Clay, silty, chalky, dark blue grey, with flint and quartz pebbles, stiff	6.5	11.5
	Clay, silty, sandy, olive grey, soft	0.3	11.8
	Clay, silty, laminated coarsely below 13.3 m with bands of chalky grey shaley clay, olive grey, firm	4.2	16.0
	Clay, shaley, many small chalk pellets, grey, stiff	0.3	16.3
	Clay, silty, many small chalk pellets, fossil fragments, blue black, stiff	2.3+	18.6

TL 41 SW 26 **4245 1118** **Eastwick Mead, Eastwick** **Block E**

Surface level +35.7 m
Water struck at +33.9 m
Shell 152 mm diameter
September 1975

Overburden 1.4 m
Mineral 2.3 m
Waste 0.9 m
Bedrock 1.3 m+

Log

Geological classification	*Lithology*	*Thickness* m	*Depth* m
	Soil	0.3	0.3
Alluvium	Clay, sandy, brown, soft	0.4	0.7
	Clay, sandy, pebbly, yellow brown, soft	0.7	1.4
	Gravel Sand: medium with coarse and some fine, predominantly quartz, yellow brown Gravel: fine with coarse, subangular to rounded flint, with a trace of rounded quartz and quartzite, and rare chalk and sandstone	2.3	3.7
	Silt, laminated with carbonaceous partings, grey, soft	0.3	4.0
	'Very clayey' sandy, silty gravel, dark grey, with a trace of chalk	0.6	4.6
Woolwich and Reading Beds	Clay, silty, mottled brown, blue, green, grey and red, firm to stiff, waxy	1.3+	5.9

Grading

Mean for deposit *percentages*			Depth below surface (m)	*percentages*					
Fines	Sand	Gravel		Fines	Sand			Gravel	
				$-\frac{1}{16}$	$+\frac{1}{16}-\frac{1}{4}$	$+\frac{1}{4}-1$	+1–4	+4–16	+16
6	37	57	1.4–2.4	12	7	20	13	27	21
			2.4–3.7	2	3	19	13	36	27
			Mean	6	5	19	13	32	25

Composition

Depth below surface (m)	Percentages by weight in + 4–16 mm fraction					
	Chalk	Flint	Quartz and Quartzite	Fossil debris	Sandstone	Others
1.4–2.4	2	76	12	–	1	9
2.4–3.7	2	89	7	–	–	2
Mean	2	83	9	–	1	5

TL 41 SW 27 **4239 1034** **Roydon Lea Farm, Harlow** **Block B**

Surface level +57.5 m
Water not struck
Shell 152 mm diameter
September 1975

Waste 7.1 m
Bedrock 0.4 m+

Log

Geological classification	*Lithology*	*Thickness* m	*Depth* m
	Soil	0.3	0.3
Head **Gravel**	Clay, sandy, pebbly, brown, soft	0.2	0.5
	'Clayey' sandy gravel Sand: medium with some fine and coarse, predominantly rounded quartz, with some angular flint, yellow brown Gravel: fine with some coarse, angular to subrounded flint, with some rounded chalk	0.9	1.4
Boulder Clay	Clay, sandy, chalky, pebbly, brown, firm	2.1	3.5
	Clay, chalky, pebbly, grey blue, stiff	2.2	5.7
	Clay, sandy, silty, pebbly, dark grey, firm	1.4	7.1
London Clay	Clay, silty, dark blue grey, stiff	0.4+	7.5

Grading

Mean for deposit *percentages*			Depth below surface (m)	*percentages*					
Fines	Sand	Gravel		Fines	Sand			Gravel	
				$-\frac{1}{16}$	$+\frac{1}{16}-\frac{1}{4}$	$+\frac{1}{4}-1$	+1–4	+4–16	+16
19	56	25	0.5–1.4	19	13	32	11	18	7

TL 41 SW 28 **4388 1437** **Overhall Farm, Gilston** **Block C**

Surface level +74.6 m
Water struck at +51.8 m
Shell 152 mm diameter
November–December 1975

Overburden 15.0 m
Mineral 8.5 m+

Log

Geological classification	*Lithology*	*Thickness* m	*Depth* m
	Soil	0.1	0.1
Boulder Clay	Clay, chalky, silty, pebbly, brown, soft	3.9	4.0
	Clay, very silty, chalky, grey becoming blue grey, hard	10.8	14.8
	Clay, silty, sandy, pebbly, brown, firm	0.2	15.0
Glacial Sand and Gravel	'Clayey' sandy gravel, chalky in upper 5.0 m Sand: medium with fine and coarse, predominantly rounded quartz, with coarse angular flint Gravel: fine with coarse, angular to subrounded white, brown and black flint, with some rounded quartz and quartzite, with a trace of sandstone, chalk, and rare fossil debris	8.5+	23.5

Grading

Mean for deposit *percentages*			Depth below surface (m)	*percentages*					
Fines	Sand	Gravel		Fines	Sand			Gravel	
				$-\frac{1}{16}$	$+\frac{1}{16}-\frac{1}{4}$	$+\frac{1}{4}-1$	+1–4	+4–16	+16
15	51	34	15.0–16.0	16	13	28	9	20	14
			16.0–17.0	13	12	38	10	20	7
			17.0–18.0	11	13	36	10	21	9
			18.0–19.0	14	17	38	8	16	7
			19.0–20.0	25	19	21	11	19	5
			20.0–21.0	18	15	22	10	21	14
			21.0–22.0	15	11	22	13	29	10
			22.0–23.0	10	10	19	15	29	17
			23.0–23.5*	2	7	18	11	23	39
			Mean	15	13	27	11	22	12

Composition

Depth below surface (m)	Percentages by weight in + 4–16 mm fraction					
	Chalk	Flint	Quartz and Quartzite	Fossil debris	Sandstone	Others
15.0–16.0	11	60	16	1	9	3
16.0–17.0	8	63	14	1	11	3
17.0–18.0	6	68	16	1	6	3
18.0–19.0	10	59	21	1	6	3
19.0–20.0	3	74	13	–	5	5
20.0–21.0	–	79	16	–	4	1
21.0–22.0	–	83	13	–	4	–
22.0–23.0	–	80	12	–	6	2
23.0–23.5	–	78	13	–	4	5
Mean	4	71	15	1	6	3

TL 41 SW 29 **4324 1308** **Eastwick Hall Farm, Eastwick** **Block C**

Surface level +67.5 m
Water struck at +62.5 m
Shell 152 mm diameter
November 1975

Overburden 10.1 m
Mineral 4.6 m
Bedrock 0.7 m+

Log

Geological classification	*Lithology*	*Thickness* m	*Depth* m
	Soil	0.3	0.3
Boulder Clay	Clay, silty, chalky, brown, soft to firm	4.7	5.0
	'Very clayey' sandy gravel Sand: fine to coarse, yellow Gravel: fine with coarse, angular flint and rounded chalk, with a trace of quartz, quartzite, sandstone and fossil debris	0.6	5.6
	Clay, silty, chalky, sandy in parts, flinty, blue, firm	3.7	9.3
	Clay, silty, pebbly, sandy, chalky, brown, firm	0.8	10.1
Glacial Sand and Gravel	'Clayey' sandy gravel Sand: medium with fine and coarse, predominantly quartz, with some angular coarse flint, yellow brown Gravel: fine with coarse, subangular to subrounded flint, with some rounded quartz and quartzite, and a trace of sandstone	4.6	14.7
London Clay	Clay, silty, sandy, brown mottled yellow brown, soft	0.2	14.9
	Clay, silty, drab olive grey, firm	0.5+	15.4

Grading

Mean for deposit *percentages*			Depth below surface (m)	*percentages*					
Fines	Sand	Gravel		Fines	Sand			Gravel	
				$-\frac{1}{16}$	$+\frac{1}{16}-\frac{1}{4}$	$+\frac{1}{4}-1$	+1–4	+4–16	+16
13	56	31	10.1–11.1	12	14	30	12	24	8
			11.1–12.1	10	20	44	9	14	3
			12.1–13.1	15	13	20	14	26	12
			13.1–14.1	11	14	25	15	26	9
			14.1–14.7	17	13	18	16	25	9
			Mean	13	15	28	13	23	8

Composition

Depth below surface (m)	Percentages by weight in + 4–16 mm fraction				
	Chalk	Flint	Quartz and Quartzite	Fossil debris	Sandstone	Others
10.1–11.1	–	78	12	–	8	2
11.1–12.1	–	80	12	–	6	2
12.1–13.1	–	78	14	–	5	3
13.1–14.1	–	78	14	–	6	2
14.1–14.7	–	80	13	–	4	3
Mean	–	79	13	–	6	2

TL 41 SW 30 4385 1240 Gilston Park, Gilston **Block C**

Surface level +61.4 m
Water not struck
Shell 152 mm diameter
November 1975

Overburden 8.2 m
Mineral 2.3 m
Waste 1.3 m
Mineral 5.4 m
Bedrock 1.8 m+

Log

Geological classification	*Lithology*	*Thickness* m	*Depth* m
	Soil	0.2	0.2
Boulder Clay	Clay, chalky, pebbly, brown to light grey, soft	3.7	3.9
	Clay, chalky, silty, pebbly, grey blue, firm	0.7	4.6
	Sand, clayey, firm	0.2	4.8
	Clay, silty, chalky, flinty, grey blue, firm	3.2	8.0
	Clay, chalky, pebbly, sandy, silty, brown, firm	0.2	8.2
Glacial Sand and Gravel	**a** Sand, clayey and pebbly at base Sand: medium with some fine and a trace of coarse, predominantly quartz, yellow brown Gravel: fine and coarse, angular to subrounded flint, with rounded chalk and some quartz, quartzite and sandstone	2.3	10.5
Boulder Clay	Clay, sandy, silty, chalky, pebbly, brown, firm	0.4	10.9
	Clay, chalky, pebbly, silty, blue grey, firm	0.6	11.5
	Clay, silty, pebbly, sandy and chalky in parts, brown, firm	0.3	11.8
Glacial Sand and Gravel	**b** Gravel, clayey at top becoming coarser with depth Sand: medium with coarse and fine, predominantly quartz, with some angular coarse flint, brown Gravel: fine and coarse with a trace of cobble, angular to subrounded flint, with some quartz and quartzite, and a trace of sandstone	5.4	17.2
London Clay (Basement Bed)	Clay, sandy, silty, olive grey to blue grey, with selenite crystals, pyrite nodules, glauconite, wood fragments and decalcified shell debris, firm	1.8+	19.0

Grading

	Mean for deposit *percentages*			Depth below surface (m)	*percentages*						
	Fines	Sand	Gravel		Fines	Sand			Gravel		
					$-\frac{1}{16}$	$+\frac{1}{16}-\frac{1}{4}$	$+\frac{1}{4}-1$	+1–4	+4–16	+16–64	+64
a	8	89	3	8.2–9.2	7	27	63	1	1	1	–
				9.2–10.2	8	18	71	2	1	–	–
				10.2–10.5	14	29	36	5	9	7	–
				Mean	8	24	63	2	2	1	–
b	7	38	55	11.8–12.8	11	12	25	10	24	18	–
				12.8–13.8	11	11	22	11	23	22	–
				13.8–14.8	11	12	14	10	26	23	4
				14.8–15.8	3	8	15	14	33	27	–
				15.8–17.2	2	4	11	14	32	37	–
				Mean	7	9	17	12	28	26	1

Composition

	Depth below surface (m)	Percentages by weight in + 4–16 mm fraction					
		Chalk	Flint	Quartz and Quartzite	Fossil debris	Sandstone	Others
a	8.2–9.2	30	37	12	–	12	9
	9.2–10.2	29	38	8	–	10	15
	10.2–10.5	16	59	11	2	6	6
	Mean	28	40	10	–	10	12
b	11.8–12.8	–	87	9	–	3	1
	12.8–13.8	–	73	19	–	5	3
	13.8–14.8	–	76	18	–	4	2
	14.8–15.8	Sample missing					
	15.8–17.2	–	77	17	–	2	4
	Mean	–	78	16	–	3	3

TL 41 SW 31 **4371 1143** **½ km SE of St. Bolophs, Harlow** **Block E**

Surface level +36.4 m
Water struck at +35.1 m
Shell 152 mm diameter
September 1975

Overburden 1.5 m
Mineral 1.9 m
Bedrock 2.6 m+

Log

Geological classification	*Lithology*	*Thickness* m	*Depth* m
	Soil	0.4	0.4
Alluvium	Clay, silty, fragments of flint, chalk and gastropod debris, brown, soft	0.6	1.0
	Clay, silty, sandy and pebbly at base, grey and brown, soft	0.5	1.5
	Gravel Sand: medium and coarse with a trace of fine, yellow Gravel: fine and coarse, angular to subangular with rare rounded flint, some rounded quartz and quartzite, and a trace of chalk and sandstone	1.9	3.4
London Clay (Basement Bed)	Sand, fine, silty, grey black, becoming brown black, with decalcified shell debris, soft	1.7	5.1
	Clay, silty, blue grey, with carbonaceous partings and shell debris, stiff	0.9+	6.0

Grading

Mean for deposit *percentages*			Depth below surface (m)	*percentages*					
Fines	Sand	Gravel		Fines	Sand			Gravel	
				$-\frac{1}{16}$	$+\frac{1}{16}-\frac{1}{4}$	$+\frac{1}{4}-1$	+1–4	+4–16	+16
3	32	65	1.5–2.5*	3	2	11	12	38	34
			2.5–3.4*	2	3	23	15	29	28
			Mean	3	2	17	13	34	31

Composition

Depth below surface (m)	Percentages by weight in + 4–16 mm fraction					
	Chalk	Flint	Quartz and Quartzite	Fossil debris	Sandstone	Others
1.5–2.5	4	86	9	–	–	1
2.5–3.4	5	76	11	–	3	5
Mean	5	81	10	–	1	3

TL 41 SW 32 **4411 1496** **Actons Farm, High Wych** **Block C**

Surface level +72.1 m
Water struck at +63.6 m
Shell (modified) 152 mm diameter
April 1976

Waste 18.6 m+

Log

Geological classification	*Lithology*	*Thickness* m	*Depth* m
	Soil	0.2	0.2
Boulder Clay	Clay, chalky, brown, soft to firm	4.1	4.3
	Clay, chalky, pebbly, dark grey, firm	4.2	8.5
	Clay, silty, very sandy, dark grey, soft	0.7	9.2
	Clay, chalky, silty, blue grey, stiff	9.4+	18.6

TL 41 SW 33 **4464 1404** **Golden Grove, Gilston** **Block C**

Surface level +69.6 m
Water struck at +63.0 m
Shell 152 mm diameter
November 1975

Overburden 2.6 m
Mineral 4.2 m
Bedrock 1.8 m+

Log

Geological classification	*Lithology*	*Thickness* m	*Depth* m
	Soil	0.2	0.2
Boulder Clay	Clay, sandy, silty, pebbly, brown, soft	2.4	2.6
Glacial Sand and Gravel	'Clayey' sandy gravel, with chalky clay band between 5.1 and 5.3 m Sand: medium with some fine and coarse, predominantly quartz, with angular coarse flint, yellow brown Gravel: fine and coarse, angular to subrounded flint, with some quartz, quartzite and sandstone, and a trace of rounded chalk	4.2	6.8
London Clay	Clay, silty, sandy, brown, soft	0.2	7.0
	Clay, sandy, silty, drab olive green, with a trace of glauconite, pyritic wood fragments and selenite crystals, firm	1.6+	8.6

Grading

Mean for deposit *percentages*			Depth below surface (m)	*percentages*					
Fines	Sand	Gravel		Fines	Sand			Gravel	
				$-\frac{1}{16}$	$+\frac{1}{16}-\frac{1}{4}$	$+\frac{1}{4}-1$	+1–4	+4–16	+16
13	53	34	2.6–3.6	28	9	20	6	18	19
			3.6–5.1	8	12	51	6	11	11
			5.3–6.3	8	10	30	11	27	14
			6.3–6.8	11	10	18	14	29	18
			Mean	13	11	34	8	19	15

Composition

Depth below surface (m)	Percentages by weight in + 4–16 mm fraction					
	Chalk	Flint	Quartz and Quartzite	Fossil debris	Sandstone	Others
2.6–3.6	2	80	12	–	4	2
3.6–5.1	1	80	10	–	7	2
5.3–6.3	2	75	13	–	8	2
6.3–6.8	–	74	17	–	7	2
Mean	1	78	12	–	7	2

TL 41 SW 34 4421 1353 Overhall Farm, Gilston **Block C**

Surface level +50.1 m
Water not struck
Shell 152 mm diameter
December 1975

Waste 1.5 m
Bedrock 2.4 m+

Log

Geological classification	*Lithology*	*Thickness* m	*Depth* m
	Soil	0.1	0.1
Head	Clay, silty, sandy, rarely pebbly, brown, soft	0.8	0.9
	Clay, silty, pebbly, yellow brown mottled red, soft	0.6	1.5
London Clay	Clay, silty, sandy in parts, blue grey, with glauconite, carbonaceous fragments and rare pyrite nodules, soft	2.4+	3.9

TL 41 SW 35 **4417 1349** **Overhall Farm, Gilston** **Block C**

Surface level +49.8 m
Water struck at +47.5 m and +26.8 m
Shell (modified) 152 mm diameter
May 1976

Waste 2.5 m
Bedrock 25.7 m+

Log

Geological classification	*Lithology*	*Thickness* m	*Depth* m
	Soil and subsoil, clay, pebbly, mid-brown	0.6	0.6
Head	Clay, silty, sandy, flint and chalk pebbles, brown, soft	0.5	1.1
	Clay, silty, very sandy, gravelly at base, brown and grey, soft	1.4	2.5
London Clay	Clay, silty, sandy, dark grey mottled brown at top, with pockets of green glauconitic fine sand, soft	0.7	3.2
	Clay, very sandy, becomes very clayey fine sand, grey green with patches of green glauconitic sand, laminated with bands of decalcified shell debris, firm	2.6	5.8
London Clay (Basement Bed)	Silt and fine sand, grey green, very hard	0.1	5.9
	Sand, fine, clayey, silty, dark grey green, with comminuted shells and race nodules, pyritic and glauconitic in pockets, firm	4.6	10.5
Woolwich and Reading Beds	Clay, silty, pebbly, mottled green grey and red, packed with broken shells and flint pebbles	0.3	10.8
	Clay, silty, mottled grey green, red and pale cream, with some fine flint pebbles	0.4	11.2
	Clay, pale grey blue streaked green and brown, with large race nodules up to 2 cm diameter	0.6	11.8
	Sand, clayey, pale grey blue	0.5	12.3
	Clay, silty, becomes sandy, grey mottled red and green becomes red brown mottled grey and light blue	2.3	14.6
	Clay, silty, fine sandy, brown streaked grey green, waxy	1.7	16.3
	Clay, silty, brown mottled grey blue, with abundant race nodules up to 1 cm diameter, waxy	0.5	16.8
	Clay, red brown mottled blue grey, stiff, waxy	0.2	17.0
	Sand, fine, clayey, dark brown, micaceous	0.2	17.2
	Sand, fine, dark brown, micaceous	1.0	18.2
	Sand, fine, clayey, dark brown mottled grey green and red, micaceous	0.7	18.9
	Sand, fine, clayey, brown mottled red, micaceous	0.3	19.2
	Sand, fine, dark brown mottled red and grey green, with many fine gravel sized pebbles of rounded black flint and a trace of rounded white quartz	0.1	19.3
	Sand, fine, increasingly clayey, dark brown mottled blue, micaceous	0.3	19.6
	Sand, fine, and clay, olive green mottled brown and red, with rounded fine black flints	0.9	20.5
Thanet Beds (Thanet Sand)	Sand, clayey, blue green, micaceous	1.0	21.5
	Sand, fine to medium, clayey, blue green mottled grey green, grey and orange brown, glauconitic	3.2	24.7
	Sand, fine, clayey, pellets of clay. Grey green becomes purple grey and green from 27.0 m, dark brown and green glauconitic burrows, laminated, micaceous	3.0	27.7
(Bullhead Bed)	Clay, silty, bright olive green, with large angular to subrounded black flints, glauconitic, firm	0.1	27.8
	Sand, fine, bright olive green, laminated, glauconitic with chalk putty and well rounded black flints	0.1	27.9
Upper Chalk	Chalk, soft puggy, white, with angular black and blue flints	0.3+	28.2

[Summary of this log published in IGS Boreholes 1976. *Rep. Inst. Geol. Sci.*, 77/10, p. 9.]

TL 41 SW 36 4472 1238 Pye Corner, Eastwick Block C

Surface level +42.7 m
Water struck at +40.0 m
Shell 152 mm diameter
November 1975

Waste 4.6 m
Bedrock 1.2 m+

Log

Geological classification	*Lithology*	*Thickness* m	*Depth* m
	Soil	0.5	0.5
	Subsoil	0.2	0.7
Head	Clay, silty, chalky, yellow brown mottled grey, soft	3.0	3.7
	Peat, clayey, black, soft	0.2	3.9
	'Clayey' sandy gravel Sand: coarse, predominantly quartz, with some angular flint, dark brown Gravel: fine to coarse, rounded to well rounded flint and quartz, with a trace of quartzite and sandstone	0.7	4.6
London Clay	Clay, silty, very sandy, olive grey, with glauconite, comminuted shells, pyrite nodules and pyritised wood	1.2+	5.8

TL 41 SW 37 4435 1178 Lodge House, Eastwick **Block C**

Surface level +46.0 m
Water not struck
Shell 152 mm diameter
November 1975

Overburden 1.1 m
Mineral 3.3 m
Bedrock 1.5 m+

Log

Geological classification	*Lithology*	*Thickness* m	*Depth* m
	Soil	0.2	0.2
Head	Clay, silty, sandy, pebbly at base, yellow brown, soft	0.9	1.1
Glacial Sand and Gravel	Gravel, 'clayey' at top becomes coarser with depth Sand: medium and coarse with some fine, predominantly quartz, with some angular coarse flint, ochre brown Gravel: coarse and fine, angular to well rounded flint, with some rounded to subrounded quartz, quartzite and sandstone, and a trace of ironstone	3.3	4.4
London Clay	Clay, silty, drab olive grey, broken shells, pyrite nodules and glauconite, soft becoming stiff	1.5+	5.9

Grading

Mean for deposit *percentages*			Depth below surface (m)	*percentages*					
Fines	Sand	Gravel		Fines	Sand			Gravel	
				$-\frac{1}{16}$	$+\frac{1}{16}-\frac{1}{4}$	$+\frac{1}{4}-1$	+1–4	+4–16	+16
6	32	62	1.1–2.1	14	6	14	11	28	27
			2.1–3.1	3	4	14	15	30	34
			3.1–4.4	2	4	14	14	25	41
			Mean	6	5	14	13	27	35

Composition

Depth below surface (m)	Percentages by weight in + 4–16 mm fraction					
	Chalk	Flint	Quartz and Quartzite	Fossil debris	Sandstone	Others
1.1–2.1	–	80	12	–	4	4
2.1–3.1	–	73	15	–	4	8
3.1–4.4	–	69	16	–	7	8
Mean	–	74	14	–	5	7

TL 41 SE 34 **4543 1464** **Mobletts, High Wych** **Block C**

Surface level +68.5 m
Water struck at +57.7 m and +52.8 m
Shell (modified) 152 mm diameter
April 1976

Waste 16.6 m
Bedrock 0.8 m+

Log

Geological classification	*Lithology*	*Thickness* m	*Depth* m
	Soil	0.2	0.2
Boulder Clay	Clay, chalky, silty, brown to grey, soft to firm	4.1	4.3
	Clay, silty, chalky, blue grey, stiff	4.2	8.5
	Clay, silty, chalky, dark blue grey, stiff	1.1	9.6
	Clay, chalky, silty, pebbly, brown, firm	0.2	9.8
	Clay, very chalky, silty, sandy, brown, firm	0.9	10.7
Glacial Sand and Gravel	**a** Pebbly sand Sand: medium with some fine and coarse, predominantly quartz Gravel: fine with some coarse, subangular to rounded flint, with quartz and quartzite, and some chalk and sandstone	1.5	12.2
Boulder Clay	Clay, chalky, silty, pebbly, blue grey, firm to stiff	1.9	14.1
	Clay, chalky, pebbly, silty, sandy, brown, firm	0.1	14.2
Glacial Sand and Gravel	**b** Gravel, becomes coarse with depth Sand: medium with coarse and some fine, predominantly rounded quartz Gravel: fine with coarse, angular to subrounded flint, with some rounded quartz and quartzite, and a trace of sandstone	2.4	16.6
London Clay	Clay, silty, fine sandy, brown, faintly laminated, firm	0.3	16.9
	Clay, silty, becomes clayey, silty fine sand, drab olive green with glauconite, wood fragments, pyrite nodules and shell debris, stiff	0.5+	17.4

Grading

	Mean for deposit *percentages*			Depth below surface (m)	*percentages*					
	Fines	Sand	Gravel		Fines	Sand			Gravel	
					$-\frac{1}{16}$	$+\frac{1}{16}-\frac{1}{4}$	$+\frac{1}{4}-1$	+1–4	+4–16	+16
a	6	71	23	10.7–11.7*	4	8	54	11	17	6
				11.7–12.2*	10	20	40	7	17	6
				Mean	6	12	49	10	17	6
b	4	38	58	14.2–15.2	5	6	17	15	37	20
				15.2–16.2*	4	4	21	13	32	26
				16.2–16.6*	2	4	24	9	21	40
				Mean	4	5	20	13	32	26

Composition

	Depth below surface (m)	Percentages by weight in + 4–16 mm fraction					
		Chalk	Flint	Quartz and Quartzite	Fossil debris	Sandstone	Others
a	10.7–11.7	3	63	17	–	9	8
	11.7–12.2	15	65	14	–	5	1
	Mean	7	64	16	–	8	5
b	14.2–15.2	–	75	18	–	3	4
	15.2–16.2	–	71	22	–	4	3
	16.2–16.6	–	62	23	–	8	7
	Mean	–	71	21	–	4	4

TL 41 SE 35 **4553 1361** **Sayes Park Farm, High Wych** **Block C**

Surface level +54.3 m
Water struck at +48.3 m
Shell 152 mm diameter
November 1975

Overburden 2.9 m
Mineral 6.9 m
Bedrock 3.8 m+

Log

Geological classification	*Lithology*	*Thickness* m	*Depth* m
	Soil	0.2	0.2
Boulder Clay	Clay, silty, brown, stiff	1.5	1.7
	Clay, silty, chalky, flinty, yellow brown, firm	1.2	2.9
Glacial Sand and Gravel	Sandy gravel, sand between 4.4 and 5.6 m Sand: medium with coarse and some fine, predominantly rounded quartz, with some angular coarse flint, ochre brown and yellow Gravel: fine and coarse, angular to subrounded flint, with rounded quartz, quartzite and some sandstone, with a trace of chalk and ironstone	6.9	9.8
London Clay	Clay, sandy, silty, brown mottled ochre brown, firm	0.3	10.1
	Clay, silty, rarely sandy, olive green to blue, with many selenite crystals and pyrite nodules, and glauconite in patches, firm	3.5+	13.6

Grading

Mean for deposit percentages			Depth below surface (m)	*percentages*					
Fines	Sand	Gravel		Fines	Sand			Gravel	
				$-\frac{1}{16}$	$+\frac{1}{16}-\frac{1}{4}$	$+\frac{1}{4}-1$	+1–4	+4–16	+16
5	49	46	2.9–3.9	10	7	17	12	32	22
			3.9–5.0	6	15	49	7	13	10
			5.0–6.0	7	9	33	9	22	20
			6.0–7.0*	3	6	28	15	31	17
			7.0–8.0*	2	5	30	18	27	18
			8.0–9.0*	3	3	11	19	35	29
			9.0–9.8*	2	6	16	21	30	25
			Mean	5	7	27	15	26	20

Composition

Depth below surface (m)	Percentages by weight in + 4–16 mm fraction					
	Chalk	Flint	Quartz and Quartzite	Fossil debris	Sandstone	Others
2.9–3.9	Sample missing					
3.9–5.0	–	80	11	–	6	3
5.0–6.0	–	70	14	–	9	7
6.0–7.0	1	76	14	–	4	5
7.0–8.0	1½	80	11	–	6	11½
8.0–9.0	–	82	10	–	4	4
9.0–9.8	–	73	18	–	5	4
Mean	1	77	13	–	5	4

TL 41 SE 36 4570 1227 Hollingson Mead, Sawbridgeworth **Block E**

Surface level +38.8 m
Water struck at +37.3 m
Shell 152 mm diameter
November 1975

Overburden 1.5 m
Mineral 4.5 m
Bedrock 3.6 m+

Log

Geological classification	*Lithology*	*Thickness* m	*Depth* m
	Soil	0.4	0.4
Alluvium	Clay, silty, peaty, brown, soft	0.5	0.9
	Clay, silty, brown, soft	0.6	1.5
	Gravel Sand: medium and coarse with some fine, predominantly rounded quartz, with some angular coarse flint Gravel: fine and coarse, angular to subrounded black flints, with some rounded quartz, quartzite, and sandstone, with a trace of chalk	4.5	6.0
London Clay	Clay, silty, grey, with carbonaceous partings and shell debris, soft	3.2	9.2
	Clay, sandy, silty, mottled grey and dark grey, firm	0.4+	9.6

Grading

Mean for deposit *percentages*			Depth below surface (m)	*percentages*					
Fines	Sand	Gravel		Fines	Sand			Gravel	
				$-\frac{1}{16}$	$+\frac{1}{16}-\frac{1}{4}$	$+\frac{1}{4}-1$	+1–4	+4–16	+16
6	37	57	1.5–2.5	9	5	11	11	34	30
			2.5–3.5	4	5	18	14	33	26
			3.5–6.0*	6	5	19	16	29	25
			Mean	6	5	17	15	31	26

Composition

Depth below surface (m)	Percentages by weight in +4–16 mm fraction					
	Chalk	Flint	Quartz and Quartzite	Fossil debris	Sandstone	Others
1.5–2.5	9	70	7	1	9	4
2.5–3.5	4	79	8	–	6	3
3.5–6.0	5	80	8	–	5	2
Mean	6	78	8	–	6	2

TL 41 SE 37 4667 1431 A. 414, High Wych **Block D**

Surface level +78.6 m
Water not struck
Shell 152 mm diameter
October 1975

Waste 20.0 m+

Log

Geological classification	*Lithology*	*Thickness* m	*Depth* m
	Soil	0.1	0.1
Boulder Clay	Clay, chalky, silty, sandy, very flinty, yellow brown, firm	1.3	1.4
	Clay, silty, chalky, brown mottled grey, with some flint pebbles, soft	1.5	2.9
	Clay, silty, chalky, flinty, blue grey, stiff	2.6	5.5
	Clay, sandy, silty, blue, very soft	2.5	8.0
	Clay, silty, chalky, dark blue, stiff to hard	10.0	18.0
	Clay, silty, chalky, blue grey, very hard, shaley	2.0+	20.0

TL 41 SE 38 4660 1338 Rederick Lane, Sawbridgeworth **Block D**

Surface level +62.4 m
Water struck at +52.3 m
Shell 152 mm diameter
April 1976

Waste 18.4 m+

Log

Geological classification	*Lithology*	*Thickness* m	*Depth* m
	Soil	0.3	0.3
Boulder Clay	Clay, silty, chalky, flinty, brown, with silt bands, soft	3.7	4.0
	Clay, chalky, silty, dark grey, soft becoming hard	2.9	6.9
	Clay, chalky, silty, olive-grey, stiff	3.2	10.1
	Clay, sandy, chalky, grey, very soft	1.3	11.4
	Clay, very chalky, silty, olive grey, soft becoming stiff	7.0+	18.4

TL 41 SE 39 4756 1446 Rivers Nursery, Sawbridgeworth **Block D**

Surface level +64.9 m
Water struck at +56.9 m and +47.4 m
Shell 152 mm diameter
March 1976

Overburden 1.0 m
Mineral 8.0 m
Waste 10.4 m
Bedrock 0.3 m+

Log

Geological classification	*Lithology*	*Thickness* m	*Depth* m
	Soil	0.4	0.4
Boulder Clay	Clay, sandy, silty, chalky, pale orange, firm	0.6	1.0
Glacial Sand and Gravel	**a** 'Clayey' pebbly sand, clayey sand grading to gravel Sand: medium and fine with a trace of coarse, predominantly rounded quartz, with a trace of angular coarse flint Gravel: fine and coarse, angular to rounded flint, with some chalk, and a trace of rounded quartz, quartzite and sandstone	8.0	9.0
Boulder Clay	Clay, silty, sandy, chalky, light brown, firm to stiff	0.9	9.9
	Clay, silty, chalky, grey becoming dark grey, stiff	7.5	17.4
	Clay, pebbly, very sandy, orange brown, firm	0.1	17.5
Glacial Sand and Gravel	**b** Sandy gravel Sand: medium with coarse and a trace of fine, predominantly quartz, with angular coarse flint Gravel: fine with coarse, subrounded to angular flint, with some rounded quartz and quartzite, with a trace of rounded sandstone and chalk	1.9	19.4
London Clay	Clay, silty, brown, firm	0.1	19.5
	Clay, silty, olive grey, firm	0.2+	19.7

Grading

	Mean for deposit *percentages*			Depth below surface (m)	*percentages*					
	Fines	Sand	Gravel		Fines	Sand			Gravel	
					$-\frac{1}{16}$	$+\frac{1}{16}-\frac{1}{4}$	$+\frac{1}{4}-1$	+1–4	+4–16	+16
a	16	69	15	1.0–2.0	24	29	32	9	6	–
				2.0–6.0	14	40	40	4	2	–
				6.0–7.0	30	29	29	3	7	2
				7.0–8.0	16	10	26	9	24	15
				8.0–9.0*	4	4	28	10	20	34
				Mean	16	29	34	6	8	7
b	4	53	43	17.5–19.4*	4	6	29	18	25	18

Composition

	Depth below surface (m)	Percentages by weight in + 4–16 mm fraction					
		Chalk	Flint	Quartz and Quartzite	Fossil debris	Sandstone	Others
a	1.0–2.0	28	63	2	½	6	½
	2.0–6.0	28	63	3½	–	3½	2
	6.0–7.0	6	70	19	–	3	2
	7.0–8.0	1	76	17	–	2	4
	8.0–9.0	9	47	21	1	12	10
	Mean	19	64	9	–	5	3
b	17.5–19.4	1	74	15	–	8	2

TL 41 SE 40 4790 1302 ½ km N. of Harlowbury, Harlow **Block E**

Surface level +41.8 m
Water struck at +40.3 m
Shell 152 mm diameter
March–April 1976

Overburden 1.5 m
Mineral 5.9 m
Waste 7.8 m
Mineral 9.2 m+

Log

Geological classification	*Lithology*	*Thickness* m	*Depth* m
	Soil	0.2	0.2
Alluvium	Clay, sandy, silty, mottled orange and brown, soft	0.4	0.6
	Peat, dark brown, soft	0.5	1.1
	Clay, olive grey, soft	0.4	1.5
Buried Channel Deposits (Glacial Sand and Gravel)	**a** Sandy gravel, clayey at top and base, pebbly, medium sand between 4.0 and 6.0 m Sand: medium with some coarse and a trace of fine, pale brown Gravel: fine with coarse, angular flint, with rounded chalk, with some rounded sandstone, quartz and quartzite, and a trace of fossil debris	5.9	7.4
	Silt, medium to coarse, sandy, olive grey to grey, soft	7.8	15.2
	b Sandy gravel, very sandy at top Sand: medium and coarse with a trace of fine, yellow brown Gravel: fine and some coarse, angular to subrounded flint, with some rounded quartz, quartzite, chalk and sandstone, and a trace of fossil debris	9.2+	24.4

Grading

	Mean for deposit *percentages*			Depth below surface (m)	*percentages*					
	Fines	Sand	Gravel		Fines	Sand			Gravel	
					$-\frac{1}{16}$	$+\frac{1}{16}-\frac{1}{4}$	$+\frac{1}{4}-1$	+1–4	+4–16	+16
a	7	62	31	1.5–2.5*	11	5	16	11	27	30
				2.5–4.0*	2	4	16	18	33	27
				4.0–5.0*	6	11	61	14	7	1
				5.0–6.0*	4	10	62	16	7	1
				6.0–7.4*	10	8	47	18	13	4
				Mean	7	7	39	16	18	13
b	4	59	37	15.2–16.2*	9	22	35	15	16	3
				16.2–17.2*	4	8	15	26	36	11
				17.2–18.2*	3	10	33	19	17	18
				18.2–19.2*	3	7	30	26	19	15
				19.2–20.2*	1	3	28	26	19	23
				20.2–21.2*	1	4	16	16	15	48
				21.2–22.2*	7	9	46	15	14	9
				22.2–23.2*	2	6	32	25	27	8
				23.2–24.4*	2	6	32	25	27	8
				Mean	4	8	30	21	21	16

Composition

	Depth below surface (m)	Percentages by weight in + 4–16 mm fraction					
		Chalk	Flint	Quartz and Quartzite	Fossil debris	Sandstone	Others
a	1.5–2.5	2	84	8	–	2	4
	2.5–4.0	9	65	12	1	4	9
	4.0–5.0	45	45	1	1	6	2
	5.0–6.0	46	40	1	1	11	1
	6.0–7.4	39	38	3	3	11	6
	Mean	27	54	6	1	7	5
b	15.2–16.2	7	83	6	1	1	2
	16.2–17.2	5	81	10	–	4	–
	17.2–18.2	6	80	9	–	3	2
	18.2–19.2	8	67	10	3	9	3
	19.2–20.2	5	84	8	1	2	–
	20.2–21.2	1	89	5	1	4	–
	21.2–22.2	11	56	14	2	8	9
	22.2–23.2	7	79	9	–	4	1
	23.2–24.4	5	81	8	1	5	–
	Mean	6	78	9	1	4	2

TL 41 SE 41 **4770 1061** **Gravel Pit Farm, Matching** **Block D**

Surface level +66.3 m
Water not struck
Shell 152 mm diameter
April 1976

Waste 18.3 m+

Log

Geological classification	*Lithology*	*Thickness* m	*Depth* m
	Soil	0.3	0.3
Boulder Clay	Clay, sandy, flinty, light olive grey, mottled, soft	0.5	0.8
	Clay, chalky, silty, flinty, yellow brown, soft	2.2	3.0
	Clay silty, chalky, mid to dark grey, with some pebbles of flint, and a trace of mudstone, limestone and sandstone, soft becoming stiff	7.3	10.3
	Clay, chalky, silty, olive grey, stiff	8.0+	18.3

TL 41 SE 42 **4847 1404** **Spring Hall, Sawbridgeworth** **Block E**

Surface level +55.3 m
Water struck at +48.0 m
Shell 152 mm diameter
April 1976

Overburden 2.2 m
Mineral 1.3 m
Waste 1.8 m
Mineral 7.3 m
Waste 8.9 m+

Log

Geological classification	*Lithology*	*Thickness* m	*Depth* m
	Soil	0.5	0.5
Head	Clay, very sandy, pebbly, yellow brown, soft	1.7	2.2
Terrace 2	**a** 'Clayey' gravel Sand: medium with coarse and some fine, predominantly rounded quartz Gravel: fine and coarse, angular flint, with a trace of rounded quartz and sandstone	1.3	3.5
Buried Channel Deposits (Glacial Sand and Gravel)	Silt, laminated, cream becoming olive grey, soft	1.8	5.3
	b Pebbly sand, with a band of dark grey, chalky silty clay between 11.2 and 11.6 m Sand: medium with some fine and coarse, predominantly rounded quartz, with a trace of chalk, pale yellow Gravel: fine with some coarse, angular to rounded flint and rounded chalk, with a trace of rounded sandstone, quartz, quartzite and fossil debris	7.3	12.6
Boulder Clay	Clay, silty, chalky, brown to grey, soft becoming stiff	8.9+	21.5

Grading

	Mean for deposit *percentages*			Depth below surface (m)	*percentages*					
	Fines	Sand	Gravel		Fines	Sand			Gravel	
					$-\frac{1}{16}$	$+\frac{1}{16}-\frac{1}{4}$	$+\frac{1}{4}-1$	+1–4	+4–16	+16
a	12	41	47	2.2–3.5	12	6	22	13	27	20
b	7	72	21	5.3–6.3*	11	8	31	14	27	9
				6.3–7.3*	8	13	52	8	16	3
				7.3–8.3*	9	17	58	7	8	1
				8.3–9.3*	6	12	53	16	12	1
				9.3–10.3*	2	9	42	16	25	6
				10.3–11.2*	5	9	36	19	25	6
				11.6–12.6*	9	51	31	4	3	2
				Mean	7	17	43	12	17	4

Composition

	Depth below surface (m)	Percentages by weight in + 4–16 mm fraction					
		Chalk	Flint	Quartz and Quartzite	Fossil debris	Sandstone	Others
a	2.2–3.5	–	94	4	–	1	1
b	5.3–6.3	36	47	4	3	6	4
	6.3–7.3	35	50	3	2	6	4
	7.3–8.3	46	39	2	3	8	2
	8.3–9.3	39	50	3	2	6	4
	9.3–10.3	44	40	5	3	5	3
	10.3–11.2	44	33	3	3	9	8
	11.6–12.6	35	41	3	3	10	8
	Mean	40	43	3	3	7	4

TL 41 SE 43 **4875 1319** **Ashlands, Sheering** **Block D**

Surface level +57.3 m
Water struck at +56.7 m and +49.0 m
Shell 152 mm diameter
March 1973

Overburden 8.3 m
Mineral 5.2 m
Waste 3.7 m
Bedrock 0.3 m+

Log

Geological classification	*Lithology*	*Thickness* m	*Depth* m
	Soil	0.3	0.3
Boulder Clay	Clay, sandy, silty, chalky, yellow brown, soft to firm	4.7	5.0
	Clay, silty, chalky, grey becoming dark grey, stiff	2.6	7.6
	Clay, silty, sandy, chalky, brown, firm	0.7	8.3
Glacial Sand and Gravel	'Clayey' sandy gravel, 'very clayey' at top Sand: medium with fine and coarse, quartz, with some angular coarse flint Gravel: coarse and fine, angular to subrounded flint, with rounded chalk, and some rounded quartz, quartzite, sandstone and fossil debris	5.2	13.5
Boulder Clay	Clay, silty, sandy, chalky, olive grey, firm becoming stiff	3.7	17.2
London Clay	Clay, silty, micaceous, olive grey, with pyrite nodules, firm	0.3+	17.5

Grading

Mean for deposit *percentages*			Depth below surface (m)	*percentages*					
Fines	Sand	Gravel		Fines	Sand			Gravel	
				$-\frac{1}{16}$	$+\frac{1}{16}-\frac{1}{4}$	$+\frac{1}{4}-1$	+1–4	+4–16	+16
11	58	31	8.3–11.4*	10	15	27	15	14	19
			11.4–12.6*	20	22	24	12	12	10
			12.6–13.5*	5	10	29	18	19	19
			Mean	11	16	27	15	14	17

Composition

Depth below surface (m)	Percentages by weight in +4–16 mm fraction					
	Chalk	Flint	Quartz and Quartzite	Fossil debris	Sandstone	Others
8.3–11.4	31	49	7	4	7	3
11.4–12.6	14	75	5	1	3	2
12.6–13.5	3	80	12	–	1	4
Mean	22	60	7	3	5	3

TL 41 SE 44 **4862 1220** **Marsh Lane, Harlow** **Block D**

Surface level +61.1 m
Water not struck
Shell 152 mm diameter
April 1976

Waste 18.6 m+

Log

Geological classification	*Lithology*	*Thickness* m	*Depth* m
	Soil and Made Ground	1.0	1.0
Boulder Clay	Clay, sandy, chalky, brown, with thin sand seams, soft	1.6	2.6
	'Clayey' sandy gravel Sand: medium with fine and coarse, yellow Gravel: coarse and fine, subangular to subrounded flint, with chalk, quartz and quartzite, and some sandstone	0.9	3.5
	Clay, sandy, silty, chalky, yellow brown, firm	4.4	7.9
	Silt, clayey, laminated, yellow brown, with bands of fine chalk sand, soft	3.9	11.8
	Clay, silty, chalky, flinty, grey becoming blue grey, stiff	6.8+	18.6

Grading

Mean for deposit *percentages*			Depth below surface (m)	*percentages*					
Fines	Sand	Gravel		Fines	Sand			Gravel	
				$-\frac{1}{16}$	$+\frac{1}{16}-\frac{1}{4}$	$+\frac{1}{4}-1$	+1–4	+4–16	+16
15	48	37	2.6–3.5	15	18	23	7	17	20

Composition

Depth below surface (m)	Percentages by weight in +4–16 mm fraction					
	Chalk	Flint	Quartz and Quartzite	Fossil debris	Sandstone	Others
2.6–3.5	29	41	14	2	12	2

TL 41 SE 45 **4876 1109** **Feltimores, Matching** **Block D**

Surface level +66.6 m
Water not struck
Shell 152 mm diameter
April 1976

Waste 0.2 m
Bedrock 4.8 m+

Log

Geological classification	*Lithology*	*Thickness* m	*Depth* m
	Soil	0.2	0.2
London Clay	Clay, mid-brown to red brown, with race nodules up to cobble size, some mica flakes, selenite crystals and poorly preserved bivalves, waxy, stiff	4.8+	5.0

TL 41 SE 46 **4837 1030** **Hubbards Hall, Matching** **Block D**

Surface level +73.9 m
Water not struck
Shell 152 mm diameter
April 1976

Waste 0.3 m
Bedrock 3.0 m+

Log

Geological classification	*Lithology*	*Thickness* m	*Depth* m
	Soil	0.3	0.3
London Clay	Clay, brown to red brown, with race nodules, and selenite crystals, waxy, stiff	3.0+	3,3

TL 41 SE 47 4961 1455 Quickbury. Sheering **Block D**

Surface level +65.7 m
Water struck at +51.6 m and +48.1 m
Shell 152 mm diameter
February–March 1976

Waste 22.8 m+

Log

Geological classification	*Lithology*	*Thickness* m	*Depth* m
	Soil	0.5	0.5
Boulder Clay	Clay, silty, sandy, chalky, yellow brown, soft	3.9	4.4
	Clay, silty, chalky, grey, firm	4.6	9.0
	Clay, silty, chalky, flinty, dark grey, stiff	3.5	12.5
	Clay, silty, chalky, grey brown, with a sandy clay band at 14.1 m, stiff	5.1	17.6
Glacial Sand and Gravel	Gravel, becoming more sandy with depth Sand: medium with some coarse and a trace of fine, predominantly rounded quartz, with some angular coarse flint, brown Gravel: coarse and fine, angular to subrounded flint, with some rounded quartz and quartzite, and a trace of rounded sandstone, chalk and fossil debris	5.2+	22.8

Grading

Mean for deposit *percentages*			Depth below surface (m)	*percentages*					
Fines	Sand	Gravel		Fines	Sand			Gravel	
				$-\frac{1}{16}$	$+\frac{1}{16}-\frac{1}{4}$	$+\frac{1}{4}-1$	+1–4	+4–16	+16
2	39	59	17.6–19.0*	3	4	15	12	34	32
			19.0–20.3*	2	3	21	13	29	32
			20.3–22.8*	1	5	26	14	23	31
			Mean	2	4	22	13	27	32

Composition

Depth below surface (m)	Percentages by weight in + 4–16 mm fraction					
	Chalk	Flint	Quartz and Quartzite	Fossil debris	Sandstone	Others
17.6–19.0	–	80	18	–	2	–
19.0–20.3	1	65	19	1	10	4
20.3–22.8	–	82	16	–	1	1
Mean	–	77	17	–	4	2

TL 41 SE 48 4956 1452 Quickbury, Sheering Block D

Surface level +60.7 m
Water struck at +47.6 m
Shell 152 mm diameter
March 1976

Overburden 13.1 m
Mineral 5.7 m
Bedrock 0.5 m+

Log

Geological classification	*Lithology*	*Thickness* m	*Depth* m
	Soil	0.4	0.4
Boulder Clay	Clay, sandy, silty, chalky, light grey brown, firm	2.6	3.0
	Clay, silty, chalky, blue grey, with a silty brown medium sand band at 5.9 m, firm to stiff	10.1	13.1
Glacial Sand and Gravel	Sandy gravel Sand: medium with coarse and a trace of fine, predominantly rounded quartz, with a trace of angular coarse flint, yellow brown Gravel: fine with coarse, subangular to subrounded flint, with some rounded quartz and quartzite, and a trace of rounded sandstone, rare chalk and fossil debris	5.7	18.8
London Clay	Clay, micaceous, olive grey, firm	0.5+	19.3

Grading

Mean for deposit *percentages*			Depth below surface (m)	*percentages*					
Fines	Sand	Gravel		Fines	Sand			Gravel	
				$-\frac{1}{16}$	$+\frac{1}{16}-\frac{1}{4}$	$+\frac{1}{4}-1$	+1–4	+4–16	+16
2	61	37	13.1–14.1*	3	6	35	20	23	13
			14.1–15.1*	2	7	29	21	19	22
			15.1–16.1*	1	9	20	20	38	12
			16.1–17.1*	1	8	28	24	25	14
			17.1–18.8*	2	9	47	15	17	10
			Mean	2	8	34	19	23	14

Composition

Depth below surface (m)	Percentages by weight in + 4–16 mm fraction					
	Chalk	Flint	Quartz and Quartzite	Fossil debris	Sandstone	Others
13.1–14.1	1	67	16	1	9	6
14.1–15.1	3	65	17	–	9	6
15.1–16.1	–	87	11	–	1	1
16.1–17.1	2½	72½	15	2	4	4
17.1–18.8	1½	67	17	1	7	6½
Mean	2	71	15	1	6	5

TL 41 SE 49 4963 1351 ¼ km S.E. of Newhouse, Sheering **Block D**

Surface level +72.8 m
Water struck at +62.8 m
Shell 152 mm diameter
March 1976

Waste 23.5 m+

Log

Geological classification	*Lithology*	*Thickness* m	*Depth* m
	Soil	0.3	0.3
Boulder Clay	Clay, silty, chalky, slightly sandy, orange brown becoming brown, with two silty, chalky fine sand bands at 8.8 and 9.0 m, firm becoming stiff	9.7	10.0
Glacial Sand and Gravel	'Clayey' sandy gravel Sand: medium with fine and coarse, becoming coarser with depth, predominantly rounded quartz, with some angular coarse flint, orange brown Gravel: fine with coarse, subangular to subrounded flint, with rounded quartz, and some rounded sandstone, with a trace of chalk and fossil debris	1.8	11.8
Boulder Clay	Clay, silty, chalky, grey, with faint laminations, firm	1.2	13.0
	Clay, silty, chalky, dark grey becoming olive grey, stiff	9.8	22.8
	Sand, clayey, chalky to very chalky, olive grey, soft	0.7+	23.5

Grading

Mean for deposit *percentages*			Depth below surface (m)	*percentages*					
Fines	Sand	Gravel		Fines	Sand			Gravel	
				$-\frac{1}{16}$	$+\frac{1}{16}-\frac{1}{4}$	$+\frac{1}{4}-1$	+1–4	+4–16	+16
18	42	40	10.0–11.8*	18	12	18	12	25	15

Composition

Depth below surface (m)	Percentages by weight in +4–16 mm fraction					
	Chalk	Flint	Quartz and Quartzite	Fossil debris	Sandstone	Others
10.0–11.8	1	67	17	1	7	7

TL 41 SE 50 4922 1285 Durrington Hall, Sheering **Block D**

Surface level +63.3 m
Water not struck
Shell 152 mm diameter
April 1976

Waste 16.0 m
Bedrock 1.3 m+

Log

Geological classification	*Lithology*	*Thickness* m	*Depth* m
	Soil	0.2	0.2
Boulder Clay	Clay, chalky, silty, flinty, brown mottled grey, soft	4.3	4.5
	Clay, chalky, silty, olive grey, soft	5.5	10.0
	Clay, chalky, silty, dark blue grey, with some flints, and rare black shale, stiff	5.3	15.3
	Clay, silty, very sandy and flinty at base, trace of chalk, olive grey, firm	0.7	16.0
London Clay	Clay, silty, micaceous, olive green, stiff	1.3+	17.3

TL 41 SE 51 **4985 1089** **Hobbs Cross, Matching** **Block D**

Surface level +87.0 m
Water struck at +80.5 m and +74.7 m
Shell 152 mm diameter
April 1976

Waste 15.7 m
Bedrock 0.7 m+

Log

Geological classification	*Lithology*	*Thickness* m	*Depth* m
	Soil and Made Ground	1.1	1.1
Boulder Clay	Clay, silty, chalky, trace of angular flints, pale orange, firm	5.4	6.5
	Sand, silty, clayey, orange, with chalk and flint, soft	0.2	6.7
	Clay, silty, chalky, blue grey, stiff	3.5	10.2
	Clay, chalky, silty, yellow brown, with orange medium to coarse sand bands at 10.2 m and 11.0 m	1.1	11.3
Glacial Sand and Gravel	'Clayey' sandy gravel, very sandy in last 1.2 m Sand: medium with fine and a trace of coarse, predominantly quartz, yellow brown Gravel: fine with coarse, angular to subrounded flint, with rounded quartz, quartzite and some sandstone	2.2	13.5
	Sands and clays, bonded, silty, pale olive, blue and yellow, soft	2.2	15.7
London Clay	Clay, silty, sandy, micaceous, brown becoming olive grey, firm	0.7+	16.4

Grading

Mean for deposit *percentages*			Depth below surface (m)	*percentages*					
Fines	Sand	Gravel		Fines	Sand			Gravel	
				$-\frac{1}{16}$	$+\frac{1}{16}-\frac{1}{4}$	$+\frac{1}{4}-1$	+1–4	+4–16	+16
16	60	24	11.3–12.3*	13	9	29	10	23	16
			12.3–13.5*	19	30	34	5	7	5
			Mean	16	21	32	7	14	10

Composition

Depth below surface (m)	Percentages by weight in + 4–16 mm fraction					
	Chalk	Flint	Quartz and Quartzite	Fossil debris	Sandstone	Others
11.3–12.3	–	59	26	–	11	4
12.3–13.5	–	62	30	–	4	4
Mean	–	61	28	–	7	4

APPENDIX G

LIST OF WORKINGS

This list includes only the principal workings. Many smaller old workings are shown as worked-out or made ground on the resource map accompanying this report.

OPERATIONAL	*Grid reference*	*Deposit*
Gaston Green	495 163	Glacial Sand and Gravel
Roydon Lea	400 105	River Terrace Deposits
ABANDONED		
Hadham Towers	430 170	Glacial Sand and Gravel
Dane Bridge	442 197	Glacial Sand and Gravel
Gravel Pit Farm	478 103	Head Gravel
RECLAIMED (by backfilling)		
Twyford Bury	495 195	First Terrace
Hobbs Cross	490 104	Head Gravel
Pole Hill 1	452 123	Glacial Sand and Gravel
Pole Hill 2	458 128	Glacial Sand and Gravel

APPENDIX H
CONVERSION TABLE, METRES TO FEET (to nearest 0.5 ft)

m	ft	m	ft	m	ft	m	ft	m	ft
0.1	0.5	6.1	20	12.1	39.5	18.1	59.5	24.1	79
0.2	0.5	6.2	20.5	12.2	40	18.2	59.5	24.2	79.5
0.3	1	6.3	20.5	12.3	40.5	18.3	60	24.3	79.5
0.4	1.5	6.4	21	12.4	40.5	18.4	60.5	24.4	80
0.5	1.5	6.5	21.5	12.5	41	18.5	60.5	24.5	80.5
0.6	2	6.6	21.5	12.6	41.5	18.6	61	24.6	80.5
0.7	2.5	6.7	22	12.7	41.5	18.7	61.5	24.7	81
0.8	2.5	6.8	22.5	12.8	42	18.8	61.5	24.8	81.5
0.9	3	6.9	22.5	12.9	42.5	18.9	62	24.9	81.5
1.0	3.5	7.0	23	13.0	42.5	19.0	62.5	25.0	82
1.1	3.5	7.1	23.5	13.1	43	19.1	62.5	25.1	82.5
1.2	4	7.2	23.5	13.2	43.5	19.2	63	25.2	82.5
1.3	4.5	7.3	24	13.3	43.5	19.3	63.5	25.3	83
1.4	4.5	7.4	24.5	13.4	44	19.4	63.5	25.4	83.5
1.5	5	7.5	24.5	13.5	44.5	19.5	64	25.5	83.5
1.6	5	7.6	25	13.6	44.5	19.6	64.5	25.6	84
1.7	5.5	7.7	25.5	13.7	45	19.7	64.5	25.7	84.5
1.8	6	7.8	25.5	13.8	45.5	19.8	65	25.8	84.5
1.9	6	7.9	26	13.9	45.5	19.9	65.5	25.9	85
2.0	6.5	8.0	26	14.0	46	20.0	65.5	26.0	85.5
2.1	7	8.1	26.5	14.1	46.5	20.1	66	26.1	85.5
2.2	7	8.2	27	14.2	46.5	20.2	66.5	26.2	86
2.3	7.5	8.3	27	14.3	47	20.3	66.5	26.3	86.5
2.4	8	8.4	27.5	14.4	47	20.4	67	26.4	86.5
2.5	8	8.5	28	14.5	47.5	20.5	67.5	26.5	87
2.6	8.5	8.6	28	14.6	48	20.6	67.5	26.6	87.5
2.7	9	8.7	28.5	14.7	48	20.7	68	26.7	87.5
2.8	9	8.8	29	14.8	48.5	20.8	68	26.8	88
2.9	9.5	8.9	29	14.9	49	20.9	68.5	26.9	88.5
3.0	10	9.0	29.5	15.0	49	21.0	69	27.0	88.5
3.1	10	9.1	30	15.1	49.5	21.1	69	27.1	89
3.2	10.5	9.2	30	15.2	50	21.2	69.5	27.2	89
3.3	11	9.3	30.5	15.3	50	21.3	70	27.3	89.5
3.4	11	9.4	31	15.4	50.5	21.4	70	27.4	90
3.5	11.5	9.5	31	15.5	51	21.5	70.5	27.5	90
3.6	12	9.6	31.5	15.6	51	21.6	71	27.6	90.5
3.7	12	9.7	32	15.7	51.5	21.7	71	27.7	91
3.8	12.5	9.8	32	15.8	52	21.8	71.5	27.8	91
3.9	13	9.9	32.5	15.9	52	21.9	72	27.9	91.5
4.0	13	10.0	33	16.0	52.5	22.0	72	28.0	92
4.1	13.5	10.1	33	16.1	53	22.1	72.5	28.1	92
4.2	14	10.2	33.5	16.2	53	22.2	73	28.2	92.5
4.3	14	10.3	34	16.3	53.5	22.3	73	28.3	93
4.4	14.5	10.4	34	16.4	54	22.4	73.5	28.4	93
4.5	15	10.5	34.5	16.5	54	22.5	74	28.5	93.5
4.6	15	10.6	35	16.6	54.5	22.6	74	28.6	94
4.7	15.5	10.7	35	16.7	55	22.7	74.5	28.7	94
4.8	15.5	10.8	35.5	16.8	55	22.8	75	28.8	94.5
4.9	16	10.9	36	16.9	55.5	22.9	75	28.9	95
5.0	16.5	11.0	36	17.0	56	23.0	75.5	29.0	95
5.1	17	11.1	36.5	17.1	56	23.1	76	29.1	95.5
5.2	17	11.2	36.5	17.2	56.5	23.2	76	29.2	96
5.3	17.5	11.3	37	17.3	57	23.3	76.5	29.3	96
5.4	17.5	11.4	37.5	17.4	57	23.4	77	29.4	96.5
5.5	18	11.5	37.5	17.5	57.5	23.5	77	29.5	97
5.6	18.5	11.6	38	17.6	57.5	23.6	77.5	29.6	97
5.7	18.5	11.7	38.5	17.7	58	23.7	78	29.7	97.5
5.8	19	11.8	38.5	17.8	58.5	23.8	78	29.8	98
5.9	19.5	11.9	39	17.9	58.5	23.9	78.5	29.9	98
6.0	19.5	12.0	39.5	18.0	59	24.0	78.5	30.0	98.5

SELECTED BIBLIOGRAPHY

ALLEN, V. T. 1936. Terminology of medium-grained sediments. *Rep. Natl Res. Counc. Washington, 1935–1936. App. 1, Rep. Comm. Sedimentation,* pp. 18–47.

ARCHER, A. A. 1969. Background and problems of an assessment of sand and gravel resources in the United Kingdom. Pp. 495–508 in *Proc. 9th Commonw. Min. Metall. Congr. 1969,* Vol. 2, Mining and Petroleum Geology. (London: Institute of Mining and Metallurgy).

— 1970a. Standardisation of the size classification of naturally occurring particles. *Géotechnique,* Vol. 20, pp. 103–107.

— 1970b. Making the most of metrication. *Quarry Managers' J.,* Vol. 54, No. 6, pp. 223–237.

ATTERBERG, A. 1905. Die rationelle Klassifikation der Sande und Kiese. *Chem. Z.,* Vol. 29, pp. 195–198.

BOULTON, G. S. 1972. Modern arctic glaciers as depositional models for former ice sheets. *Q. J. Geol. Soc. London,* Vol. 128, pp. 361–393.

BRISTOW, C. R. and COX, F. C. 1973. The Gipping Till: a reappraisal of East Anglian glacial stratigraphy. *Q. J. Geol. Soc. London,* Vol. 129, pp. 1–37.

BRITISH STANDARD 1377. 1967. *Methods of testing soils for civil engineering purposes* (London: British Standards Institution) 233 pp.

BUREAU OF MINES AND GEOLOGICAL SURVEY. 1948. *Mineral Resources of the United States* (Washington D.C.: Public Affairs Press), pp. 14–17.

CLAYTON, K. M. 1957. Some aspects of the glacial deposits of Essex. *Proc. Geol. Assoc.,* Vol. 68, pp. 1–21.

— 1960. Land forms of parts of southern Essex. *Trans. Inst. Brit. Geogr.,* Vol. 28, pp. 55–74.

— and BROWN, J. C. 1958. The glacial deposits around Hertford. *Proc. Geol. Assoc.,* Vol. 69, pp. 103–119.

DAVIES, M. C., STANDON-BATT, L. and others. 1965. Records of wells in the area of New Series One-inch (Geological) Epping (240), Chelmsford (241), and Brightlingsea (242) sheets. *Water Supply Pap. Geol. Surv. Well Cat. Ser.*

HARRIS, P. M., THURRELL, R. G., HEALING, R. A. and ARCHER, A. A. 1974. Aggregates in Britain. *Proc. R. Soc.,* Ser. A, Vol. 339, pp. 329–353.

HESTER, S. W. 1965. Stratigraphy and palaeogeography of the Woolwich and Reading Beds. *Bull. Geol. Surv. G.B.* No. 23, pp. 117–137.

IRVING, A. 1897. On the geology of the Stort Valley (Herts. and Essex). *Proc. Geol. Assoc.,* Vol. 15, pp. 224–237.

— and IRVING, P. A. 1913. The Harlow Boulder Clay. *Rep. Brit. Assn.,* No. 480.

LANE, E. W. and others. 1947. Report of the sub-committee on sediment terminology. *Trans. Am. Geophys. Union,* Vol. 28, pp. 936–938.

PETTIJOHN, F. J. 1957. *Sedimentary Rocks.* 2nd Edition (London: Harper and Row).

PERRIN, R. M. S., DAVIES, H. and FYSH, M. D. 1973. Lithology of the Chalky Boulder Clay. *Nat. Phys. Sci.,* Vol. 245, pp. 101–104.

PRESTWICH, J. 1850. The structure of the strata between the London Clay and the Chalk in the London and Hampshire Tertiary systems. Pt. 1. The Basement Bed of the London Clay. *Q.J. Geol. Soc. London,* Vol. 6, pp. 252–281.

— 1852. Pt. 2. The Thanet Sands. *Q.J. Geol. Soc. London,* Vol. 8, pp. 235–268.

— 1854. Pt. 3. The Woolwich and Reading Series. *Q.J. Geol. Soc. London,* Vol. 10, pp. 75–170.

ROSE, J., ALLEN, P. and HEY, R. W. 1976. Middle Pleistocene stratigraphy in southern East Anglia. *Nature,* Vol. 263, pp. 492–494.

SAYER, A. R., HARVEY, B. I. and others. 1965. Records of wells in the area of New Series One-inch (Geological) Great Dunmow (222) and Braintree (223) sheets. *Water Supply Pap. Geol. Surv. Well Cat. Ser.*

THURRELL, R. G. 1971. The assessment of mineral resources with particular reference to sand and gravel. *Quarry Managers' J.,* Vol. 55, pp. 19–25.

TWENHOFFEL, W. H. 1937. Terminology of the fine-grained mechanical sediments. *Rep. Natl Res. Counc. Washington, 1936–1937, App. 1, Rep. Comm. Sedimentation,* pp. 81–104.

UDDEN, J. A. 1914. Mechanical composition of clastic sediments. *Bull. Geol. Soc. Am.,* Vol. 25, pp. 655–744.

WARREN, S. H. 1957. On the early pebble gravels of the Thames Basin from the Hertfordshire-Essex border to Clacton-on-Sea. *Geol. Mag.,* Vol. 94, pp. 40–46.

WENTWORTH, C. K. 1922. A scale of grade and class terms for clastic sediments. *J. Geol.,* Vol. 30, pp. 377–392.

— 1935. The terminology of coarse sediments. *Bull. Natl Res. Counc. Washington,* No. 98, pp. 225–246.

WHITAKER, W. 1924. The geology of the country around Hertford. *Mem. Geol. Surv. G.B.*

WILLMAN, H. B. 1942. Geology and mineral resources of the Marseilles, Ottawa and Streator quadrangles. *Bull. Illinois State Geol. Surv.,* No. 66, pp. 343–344.

WOODLAND, A. W. 1970. The buried tunnel valleys of East Anglia. *Proc. Yorkshire Geol. Soc.,* Vol. 37, pp. 521–578.

WOOLDRIDGE, S. W. 1952. Some marginal drainage features of the Chalky Boulder Clay ice sheet in Hertfordshire. *Proc. Geol. Assoc.,* Vol. 64, pp. 208–231.

The following reports of the Institute relate particularly to bulk mineral resources

Reports of the Institute of Geological Sciences

Assessment of British Sand and Gravel Resources

1 The sand and gravel resources of the country south-east of Norwich, Norfolk: Resource sheet TG 20. E. F. P. Nickless. Report 71/20 ISBN 0 11 880216 £1.15

2 The sand and gravel resources of the country around Witham, Essex: Resource sheet TL 81. H. J. E. Haggard. Report 72/6 ISBN 0 11 880588 6 £1.20

3 The sand and gravel resources of the area south and west of Woodbridge, Suffolk: Resource sheet TM 24. R. Allender and S. E. Hollyer.
Report 72/9 ISBN 0 11 880596 7 £1.70

4 The sand and gravel resources of the country around Maldon, Essex: Resource sheet TL 80. J. D. Ambrose. Report 73/1 ISBN 0 11 880600 9 £1.20

5 The sand and gravel resources of the country around Hethersett, Norfolk: Resource sheet TG 10. E. F. P. Nickless.
Report 73/4 ISBN 0 11 880606 8 £1.60

6 The sand and gravel resources of the country around Terling, Essex: Resource sheet TL 71. C. H. Eaton. Report 73/5 ISBN 0 11 880608 4 £1.20

7 The sand and gravel resources of the country around Layer Breton and Tolleshunt D'Arcy, Essex: Resource sheet TL 91 and part of TL 90. J. D. Ambrose.
Report 73/8 ISBN 0 11 990614 9 £1.30

8 The sand and gravel resources of the country around Shotley and Felixstowe, Suffolk: Resource sheet TM 23. R. Allender and S. E. Hollyer.
Report 73/13 ISBN 0 11 880625 4 £1.60

9 The sand and gravel resources of the country around Attlebridge, Norfolk: Resource sheet TG 11. E. F. P. Nickless.
Report 73/15 ISBN 0 11 880658 0 £1.85

10 The sand and gravel resources of the country west of Colchester, Essex: Resource sheet TL 92. J. D. Ambrose. Report 74/6 ISBN 0 11 880671 8 £1.45

11 The sand and gravel resources of the country around Tattingstone, Suffolk: Resource sheet TM 13. S. E. Hollyer. Report 74/9 ISBN 0 11 880675 0 £1.95

12 The sand and gravel resources of the country around Gerrards Cross, Buckinghamshire: Resource sheets SU 99, TQ 08 and TQ 09. H. C. Squirrell.
Report 74/14 ISBN 0 11 880710 2 £2.20

13 The sand and gravel resources of the country east of Chelmsford, Essex: Resource sheet TM 70. M. R. Clarke. ISBN 0 11 880744 7 £3.50

14 The sand and gravel resources of the country east of Colchester, Essex: Resource sheet TM 02. J. D. Ambrose. ISBN 0 11 880745 5 £3.25

15 The sand and gravel resources of the country around Newton on Trent, Lincolnshire: Resource sheet SK 87. D. Price.
ISBN 0 11 880746 3 £3.00

16 The sand and gravel resources of the country around Braintree, Essex: Resource sheet TL 72. M. R. Clarke. ISBN 0 11 880747 1 £3.50

17 The sand and gravel resources of the country around Besthorpe, Nottinghamshire: Resource sheet SK 86, part SK 76. J. R. Gozzard.
ISBN 0 11 880748 X £3.00

18 The sand and gravel resources of the Thames Valley, the country around Cricklade, Wiltshire: Resource sheets SU 09/19 and parts of SP 00/10. P. R. Robson.
ISBN 0 11 880749 8 £3.00

19 The sand and gravel resources of the country south of Gainsborough, Lincolnshire: Resource sheet SK 88 and part of SK 78. J. H. Lovell.
ISBN 0 11 880750 1 £2.50

20 The sand and gravel resources of the country east of Newark upon Trent, Nottinghamshire: Resource sheet SK 85. J. R. Gozzard.
ISBN 011 880751 X £2.75

21 The sand and gravel resources of the Thames and Kennet Valleys, the country around Pangbourne, Berkshire: Resource sheet SU 67. H. C. Squirrell.
ISBN 0 11 880752 8 £3.25

22 The sand and gravel resources of the country north-west of Scunthorpe, Humberside: Resource sheet SE 81. J. W. C. James.
ISBN 0 11 880753 6 £3.00

23 The sand and gravel resources of the Thames Valley, the country between Lechlade and Standlake: Resource sheet SP 30 and parts of SP 20, SU 29 and SU 39. P. Robson.
ISBN 0 11 881252 1 £7.25

24 The sand and gravel resources of the country around Aldermaston, Berkshire: Parts of resource sheets SU 56 and SU 66. H. C. Squirrell.
ISBN 0 11 881253 X £5.00

25 The celestite resources of the area north-east of Bristol: Resource sheet ST 68 and parts of ST 59, 69, 79, 58, 78, 68 and 77. E. F. P. Nickless, S. J. Booth and P. N. Mosley.
ISBN 0 11 881262 9 £5.00

26 The limestone and dolomite resources of the country around Monyash, Derbyshire: Resource sheet SK 16. F. C. Cox and D. McC. Bridge.
ISBN 0 11 881263 7 £7.00

27 The sand and gravel resources of the country west and south of Lincoln, Lincolnshire: Resource sheets SK 95, SK 96 and SK 97. I. Jackson.
ISBN 0 11 884003 7 £6.00

28 The sand and gravel resources of the country around Eynsham, Oxfordshire: Resource sheet SP 40 and part of SP 41. W. J. R. Harries.
ISBN 0 11 884012 6 £3.00

29 The sand and gravel resources of the country south-west of Scunthorpe, Humberside: Resource sheet SE 80. J. H. Lovell.
ISBN 0 11 884013 4 £3.50

30 Procedure for the assessment of limestone resources. F. C. Cox, D. McC. Bridge and J. H. Hull.
ISBN 0 11 884030 4 £1.25

31 The sand and gravel resources of the country west of Newark upon Trent, Nottinghamshire: Resource sheet SK 75. D. Price and P. J. Rogers.
ISBN 0 11 884031 2 £3.50

32 The sand and gravel resources of the country around Sonning and Henley: Resource sheets SU 77 and SU 78. H. C. Squirrell.
ISBN 0 11 884032 0 £5.25

33 The sand and gravel resources of the country north of Gainsborough: Resource sheet SK 89. J. Gozzard and D. Price.
ISBN 0 11 884033 9 £4.50

34 The sand and gravel resources of the Dengie Peninsula, Essex: Resource sheet TL 90, etc. M. B. Simmons.
ISBN 0 11 884081 9 £5.00

35 The sand and gravel resources of the country around Darvel: Resource sheet NS53, 63, etc. E. F. P. Nickless, A. M. Aitken and A. A. McMillan.
ISBN 0 11 884082 7 £7.00

36 The sand and gravel resources of the country around Southend-on-Sea, Essex: Resource sheets TQ 78, 79 etc. S. E. Hollyer and M. B. Simmons.
ISBN 0 11 884083 5 £7.50

37 The sand and gravel resources of the country around Bawtry, South Yorkshire: Resource sheet SK 69. A. R. Clayton.
ISBN 0 11 884053 3 £5.75

38 The sand and gravel resources of the country around Abingdon, Oxfordshire: Resource sheet SU 49, 59, SP 40, 50. C. E. Corser.
ISBN 0 11 884084 5 £5.50

39 The sand and gravel resources of the Blackwater Valley (Aldershot) area: Resource sheet SU 85, 86, parts SU 84, 94, 95, 96. M. R. Clarke and A. J. Dixon.
ISBN 0 11 884085 1 £7.00

40 The sand and gravel resources of the country west of Darlington, County Durham: Resource sheet NZ 11, 21. A. Smith.
ISBN 0 11 884086 X £4.75

41 The sand and gravel resources of the country around Garmouth, Grampian Region: Resource sheet NJ 36. A. M. Aitken, J. W. Merritt and A. J. Shaw.
ISBN 0 11 884090 8 £8.75

42 The sand and gravel resources of the country around Maidenhead and Marlow: Resource sheet SU 88, parts SU 87, 97, 98. P. N. Dunkley.
ISBN 0 11 884091 6 £5.00

43 The sand and gravel resources of the country around Misterton, Nottinghamshire: Resource sheet SK 79. D. Thomas and D. Price.
ISBN 0 11 884092 4 £5.25

44 The sand and gravel resources of the country around Sedgefield, Durham: Resource sheet NZ 32. M. D. A. Samuel.
ISBN 0 11 884093 2 £5.75

45 The sand and gravel resources of the country around Brampton, Cumbria: Resource sheet NY 55, part 56. I. Jackson.
ISBN 0 11 884094 0 £6.75

46 The sand and gravel resources of the country north of Harlow, Essex: Resource sheet TL 41. P. M. Hopson.
ISBN 0 11 884107 6 £9.50

47 The limestone and dolomite resources of the country around Wirksworth, Derbyshire: Resource sheet SK 25, part 35. F. C. Cox and D. J. Harrison.
ISBN 0 11 884108 4 £9.75

Reports of the Institute of Geological Sciences

Other Reports

69/9 Sand and gravel resources of the inner Moray Firth. A. L. Harris and J. D. Peacock.
ISBN 0 11 880106 6 35p

70/4 Sands and gravels of the southern counties of Scotland. G. A. Goodlet.
ISBN 0 11 880105 8 90p

72/8 The use and resources of moulding sand in Northern Ireland. R. A. Old.
ISBN 0 11 881594 0 30p

73/9 The superficial deposits of the Firth of Clyde and its sea lochs. C. E. Deegan, R. Kirby, I. Rae and R. Floyd.
ISBN 0 11 880617 3 95p

77/1 Sources of aggregate in Northern Ireland (2nd edition). I. B. Cameron.
ISBN 0 11 881279 3 70p

77/2 Sand and gravel resources of the Grampian Region. J. D. Peacock and others.
ISBN 0 11 881282 3 80p

77/5 Sand and gravel resources of the Fife Region. M. A. E. Browne.
ISBN 0 11 884004 5 60p

77/6 Sand and gravel resources of the Tayside Region. I. B. Paterson.
ISBN 0 11 884008 8 £1.40

77/8 Sand and gravel resources of the Strathclyde Region. I. B. Cameron and others.
ISBN 0 11 884028 2 £2.50

77/9 Sand and gravel resources of the Central Region, Scotland. M. A. E. Browne.
ISBN 0 11 884016 9 £1.35

77/19 Sand and gravel resources of the Borders Region, Scotland. A. D. McAdam.
ISBN 0 11 884025 8 £1.00

77/22 Sand and gravel resources of the Dumfries and Galloway Region of Scotland. I. B. Cameron.
ISBN 0 11 884025 8 £1.20

78/1 Sand and gravel resources of the Lothian Region of Scotland. A. D. McAdam.
ISBN 0 11 884042 8 £1.00

78/8 Sand and gravel resources of the Highland Region. W. Mykura, D. L. Ross and F. May.
ISBN 0 11 884050 9 £3.00

Dd 596792 K8

Typeset for the Institute of Geological Sciences by Frowde and Co. (Printers) Limited, London SE5

Printed in England for Her Majesty's Stationery Office by Commercial Colour Press, London E7